T0222457

Organisms, Agency, and Evolution

The central insight of Darwin's *Origin of Species* is that evolution is an ecological phenomenon, arising from the activities of organisms in the 'struggle for life'. By contrast, the Modern Synthesis theory of evolution, that rose to prominence in the twentieth century, presents evolution as a fundamentally molecular phenomenon, occurring in populations of suborganismal entities – genes. After nearly a century of success, the Modern Synthesis theory is now being challenged by empirical advances in the study of organismal development and inheritance. In this important study, D.M. Walsh shows that the principal defect of the Modern Synthesis resides in its rejection of Darwin's organismal perspective, and argues for 'situated Darwinism': an alternative, organism-centred conception of evolution that prioritises organisms as adaptive agents. His book will be of interest to scholars and advanced students of evolutionary biology and the philosophy of biology.

D.M. WALSH is Professor in the Department of Philosophy, the Institute for the History and Philosophy of Science and Technology, and the Department of Ecology and Evolutionary Biology at the University of Toronto. He is the editor of *Naturalism, Evolution and Mind* (Cambridge, 2001) and the coeditor of *Evolutionary Biology: Conceptual, Ethical and Religious Issues* (with R. Paul Thompson, Cambridge, 2014).

Organisms, Agency, and Evolution

D.M. Walsh

University of Toronto

CAMBRIDGE
UNIVERSITY PRESS

CAMBRIDGE
UNIVERSITY PRESS

University Printing House, Cambridge CB2 8BS, United Kingdom

One Liberty Plaza, 20th Floor, New York, NY 10006, USA

477 Williamstown Road, Port Melbourne, VIC 3207, Australia

314-321, 3rd Floor, Plot 3, Splendor Forum, Jasola District Centre, New Delhi-110025, India

79 Anson Road, #06-04/06, Singapore 079906

Cambridge University Press is part of the University of Cambridge.

It furthers the University's mission by disseminating knowledge in the pursuit of education, learning and research at the highest international levels of excellence.

www.cambridge.org
Information on this title: www.cambridge.org/9781107552425

First published 2015
First paperback edition 2018

A catalogue record for this publication is available from the British Library

ISBN 978-1-107-12210-9 Hardback
ISBN 978-1-107-55242-5 Paperback

Cambridge University Press has no responsibility for the persistence or accuracy of URLs for external or third-party internet websites referred to in this publication, and does not guarantee that any content on such websites is, or will remain, accurate or appropriate.

For Margaret O'Malley

Contents

Preface

When I took my first ethology course, I was instructed to make an ethogram. An ethogram is a catalogue of the movements, postures and sounds of a target organism whose behaviour one wishes to study. The objective of the ethogram is to free the observation of behaviour from any taint of purpose or intention. We were to identify behaviours with manifestations, rather than motives. I was told that my Columbian Ground Squirrels (*Spermophilus columbianus*) were standing on their hind legs at full height, rather than surveying the scene for predators. They were emitting a high-pitched 'bark', rather than alerting their fellow colony members to imminent dangers. As a biddable undergraduate, I dutifully went along with this, even though I could see that my squirrels were looking for something, and warning each other. My brush with ethology taught me three things. The first is that midnight black is the best hair-dye tint for painting patterns on ground squirrels. The other two are less practical. They are that organisms are fundamentally purposive entities, and that biologists have an animadversion to purpose. These perplex me: why should the phenomenon that demarcates the domain of biology be off-limits to biology?

This is more than just an idiosyncratic bemusement on my part. It is a *leitmotif* that runs through the entire history of biology. Since its inception, biology seems to have been torn by the evidently incompatible demands of treating organisms as natural entities, like everything else, and as singularly peculiar (naturally purposive) things, unlike anything else. One of the common strategies has been to attempt to minimise the distinctiveness of organisms, to show that the problematic nature of organisms is incidental to a comprehensive understanding of understanding of biology, and that the principles by which we account for nonliving phenomena are wholly adequate to the explanation of living phenomena. This is a bold move, and it is fair to say that it has paid dividends.

Nowhere is it pursued more vigorously, or more successfully, than in the theory of evolution that descended from Darwin's and grew to prominence throughout the twentieth century. The Modern Synthesis theory of evolution is elegant and powerful. Inspired by the molecular revolution in biology, the Modern Synthesis circumvents the 'organism' issue by making genes the canonical unit of biological organisation. Modern Synthesis evolution is a fundamentally molecular phenomenon. It is the process in which a giant molecule, DNA, is replicated, and transmitted from one generation to the next, whereupon it builds entities ('vehicles' or 'interactors') that help to spread further replicates.

No one can gainsay the advances in our understanding of biology ushered in by the Modern Synthesis. It is now coming up for a century of unparalleled success. But we are beginning to see intimations that, perhaps, this disorganicised evolutionary thinking may be running up against its limitations. These deficiencies, by my reckoning, occur at just the places where our understanding of evolution could be enhanced by paying attention to the contribution of organisms *as* organisms.

Whether or not the Modern Synthesis is reaching the end of its useful life, there is ample justification for exploring an alternative. It lies in the fact that Modern Synthesis thinking misrepresents the metaphysics of evolution. Evolution, properly construed, is not so much a molecular phenomenon as an ecological one. It arises out of what organisms do in the pursuit of their ways of life. That, I take it, is the lesson of cardinal importance to be drawn from Darwin's *Origin of Species*. But it is precisely this perspective that has been obscured by the marginalisation of organisms that has taken hold under the Modern Synthesis. I think it is a salutary exercise to contemplate what we might have missed by leaving organisms out of our understanding of evolution. That is the objective of this book.

The book comprises three main parts. Part I, 'The Eclipse of the Organism', is my attempt to explore some of the reasons behind the peripheralisation of organisms in Modern Synthesis evolutionary biology. They are, as the ecologists say, 'complex and interacting'. I have chosen to tease apart (as the anatomists say), and concentrate on, three: methodological, theoretical and empirical. After an introductory chapter that attempts to motivate the problem that organisms pose for the understanding of evolution, I proceed to make the case. Chapter 1 argues that Modern Synthesis biology has embraced the precepts and procedures of mechanism. This is hardly surprising, as mechanism is the methodological standard under which the startling advance of modern sciences has marched. Be that as it may, mechanism appears to be ill-suited to the task of capturing the distinctive features of organisms. In Chapter 2, I explore the way that the specific kind of population thinking ushered in at the inception of the Modern Synthesis leaves the (erroneous) impression that there is no room for organisms in evolutionary dynamics. Our current theory of population change gives us ensembles of abstract entities – gene types – and the forces that propel them around genotype space, but there are no organisms in its ontology. One is invited to suppose that, to the extent that this theory is adequate, organisms are redundant to theoretical biology.

Chapter 3 explores the consequences of the discovery and intensive concentration on genes that has inspired the growth of twentieth-century biology (and continues to do so). The gene has encouraged what I call the 'fractionation' of evolution. Thanks to gene centrism we can think of the component processes of evolution – inheritance, development, adaptive change and the origin of novelties – as discrete and autonomous. Together, they comprise evolution. They are united by the fact that the gene is the canonical unit in the explanation of each process.

Fractionation, I think, is the linchpin of Modern Synthesis, gene-centred thinking. It is what undergirds the Modern Synthesis approach to evolution that I call Replicator Biology. If fractionation is right, I reckon, so is Replicator Biology. Fractionation amounts to the great empirical wager of the Modern Synthesis. While fractionation is not usually the explicit target of those who wish to challenge the Modern Synthesis, it is the implications of fractionation that have been most vigorously opposed in recent evolutionary biology and its philosophy.

Part II, 'Beyond Replicator Biology', explores the ways that recent advances in our understanding the processes of inheritance, development and the production of evolutionary novelties challenge the fractionated picture of evolution. Chapter 4 argues that the Modern Synthesis mischaracterises inheritance. For the purposes of understanding evolution, inheritance should be construed as a gross pattern of intergenerational resemblances and differences. Instead it has been recast as the process in which genetic material is replicated and transmitted from parent to offspring. But the process of transmission isn't adequate to capture all that is evolutionarily important about the pattern of inheritance. In its place, I offer in outline a conception I call 'inheritance holism'. Chapter 5 discusses the Modern Synthesis conviction that genes are units of phenotypic control. The appeal of this picture is lent a considerable degree of support by the much maligned 'information' or 'genetic program' metaphor. While I find the standard battery of arguments against the 'genetic information' or 'genetic program' programme inconclusive, I offer a new one that I believe successfully undermines the presumptive privilege of genes in development. Here again, I argue that genes are generally not units of phenotype control; organisms are. Chapter 6 addresses what I take to be a deep conceptual confusion concerning the relation between natural selection and organismal development in current evolutionary thinking. Typically, the process of development is pitted against the process of selection, as a competing evolutionary force. Selection, the force, is the source of adaptive bias in evolutionary change, whereas development is fundamentally conservative. This, I argue, is a misconstrual of both the process of selection and development. Chapter 7 seeks to rectify it. Rather than the replication and transmission of genes as the unifying processes in evolution, I propose that organismal development (broadly construed) is. Evolution, I maintain, is development writ large.

Part III lays the conceptual foundations of an organism-centred alternative to Modern Synthesis thinking that I call 'Situated Darwinism'. Situated Darwinism is an account of how evolution falls out as a consequence of organisms' purposeful pursuit of their ways of life. The central guiding idea is that organisms are adaptively engaged in what I call, borrowing a term from ecological psychology, their system of 'affordances'. This perspective requires a significant amount of conceptual retooling. Chapter 8 argues that the Modern Synthesis theory has comprehensively misrepresented one of its principal explananda, adaptation. Adaptation, I argue, should be seen as a response to affordances, and not to external environments. The notion of affordance crucially implies purpose. If affordances figure in evolution, then so do purposes. In Chapter 9 I claim that the principal impediment to generating a comprehensive alternative to the Modern Synthesis has been the modern antipathy to natural purpose. I argue that our post-Scientific Revolution dete-leologised worldview has conspired against the inclusion of organisms in evolution. Chapter 10 asserts that this purposive perspective cannot just be grafted onto the Modern Synthesis theory. The reason is that organisms parti-cipate in evolution as agents. In order to accommodate this fact we need a special kind of theory, that I call an 'agent theory'. The Modern Synthesis is an 'object' theory. Agents do not figure *as agents* in object theories. So the Modern Synthesis cannot simply assimilate the organismal perspective as an add-on. Chapter 11 draws together the various strands of the preceding discus-sion in order to give a more explicit expression of this alternative 'Situated Darwinism'. It compares the conceptual underpinning of this nascent view of evolution with those of the more traditional Modern Synthesis. It illustrates the various ways in which organisms, as purposive agents, contribute to – indeed *enact* – evolution. The result is a way of thinking about evolution that is somewhat at variance with the orthodox Modern Synthesis.

I make only minor apologies for not saying much in what follows about how this conception of evolution might alter the way biologists approach the empirical investigation of evolution. I wish I could say more, but I feel both that I can't, and that even if I could I shouldn't. I have refrained in part because my interests lie more in understanding the metaphysics of evolution – what happens when evolution happens – than in how to study what happens when evolution happens. Mostly, I have demurred because it is presumptuous for a nonpracticing biologist to tell biologists how to go about their business. I have neither the expertise nor the hubris even to attempt that. I do, however, find myself sufficiently in awe of the ingenuity of practicing scientists to believe that if this way of thinking about evolution is viable and calls for a new way of studying it, biologists will be able to devise appropriate methods and proce-dures. My guess, if I were to venture one, would be that less change needs to be made to biologists' practices than to our interpretations of their findings.

Nor do I say much in what follows about the individuation of organisms. This, too, is a difficult issue, one that has generated an impressive philosophical literature, but not a great deal of consensus. There may be no general agreement on what an organism is, but there is, nevertheless, a reasonable degree of agreement on which things are organisms. That is sufficient for my purposes. All organisms, from the most rudimentary to the most complex, manifestly possess the capacities – plasticity, robustness, purposiveness – that, I maintain, make them agents of evolution.

I should say a word about my intended audience: there isn't one. (That may turn out to be prescient.) I would hope that this material is of interest to philosophers and evolutionary biologists alike, and even to historians of science. I think these considerations may also be relevant to those working in the philosophies of action and mind, and in naturalised meta-ethics. In what follows I haven't presupposed any particular expertise in any of these disciplines, in the hope that it may be accessible to practitioners in all these fields.

I started formulating the view I present here some time ago, during a short sabbatical spent as a guest of Richard Lewontin at Harvard University. I am happy to acknowledge my deep debts to him. One of my motivations in preparing this work has been to try to come to grips with Lewontin's repeated call for biologists and philosophers to revise the way we think of organisms in relation to their environments. This is the kernel of my enjoinder that we should understand organisms as agents embedded in a system of affordances. I hope I have made some progress toward the revision he has in mind.

I was delighted when Michael Ruse invited me to commit my perplexity to paper for his Cambridge Philosophy of Biology series. Little was he to know how deep my puzzlement ran or how widely it would ramify, nor was I for that matter. The exercise has been an education for me. The result is a little bit late to say the least, but here it is. I am extremely grateful to Michael for his steadfast encouragement, even through long stretches of time in which it looked as though no manuscript would ever materialise. I am also grateful to Hilary Gaskin at Cambridge University Press for her forbearance.

The old proverb says it takes a village to write a book. I'm privileged that my village is peopled by such fine, gentle folk, and really smart too – in fact, I usually feel like its idiot. It comprises teachers, students, colleagues and family. I acknowledge their help and support, and I am greatly obliged to them all. Richard C. Fox, Robert L. Carroll, Mark Sainsbury, David Papineau and Farish Jenkins have all taught me with patience and kindness. I reserve special thanks in this regard for Elliott Sober. My own move from biology to philosophy was largely inspired by Elliott. He later became my mentor in the field. My debt to him is inestimable. He is not to be held responsible, of course, for my having gone off course.

I have had the honour of working with wonderful students: Rachel Bryant, Michael Cournoyea, Alex Djedovic, Eugene Earnshaw-Whyte, Fermin Fulda,

Cory Lewis, Parisa Moosavi. It is a terrific thrill to be challenged by such acute minds. They have all helped me to refine and extend my thinking in myriad ways. Mike Stuart taught me about vitalism. Cory Lewis and Alex Djedovic each read through the entire penultimate draft of the entire manuscript and offered me extremely helpful advice. Without their astute critiques the final result would be, well, a lot worse. I have had terrific colleagues over the years who have given me help, support and inspiration. I'm indebted to Alexander Bird, Anjan Chakravartty, Philippe Huneman, Mohan Matthen, Mark Solovey, Larry Shapiro, Jacob Stegenga and Marga Vicedo. I should single out André Ariew, from whom I have learned so much. And there is family. Deborah Kohn has given her loyalty and love selflessly, even to the extent of surrendering her comfy wicker chair to the cause of philosophy. Aoiffe and Lia are both both a wonder and a joy. There are even mothers: Margaret Mary O'Malley and Dr V.L. Kohn. Sadly, since this manuscript was submitted for publication, Dr Kohn died. I do wish to express my admiration for her, and acknowledge my debt to her.

I began writing this manuscript while a visiting scholar at the KLI institute for the Study for Natural Complex Systems. In fact most of the first version was written during various sojourns there over the years (first in Altenberg, then in Klösterneuburg, Austria). I am extremely grateful to my friends at the KLI for offering such a convivial and productive atmosphere. I would particularly like to mention Gerd Müller, Werner Callebaut, Isabella Sarto-Jackson and Eva Kärner in this regard. Sadly, our community lost Werner Callebaut, again during the short time between the submission of this manuscript and its publication. The loss to philosophy of biology is difficult to fathom. I am grateful to LaMancha Hub in the Scottish Borders and the public libraries in Toronto; Chagrin Falls, Ohio; and Ellsworth, Maine, for providing me with spaces to work. I am particularly grateful to David Depew and Jonathan Kaplan, both of whom read the penultimate version of the manuscript and provided wonderfully insightful comments.

Despite the efforts of all these good people, inaccuracies, infelicities, incomprehensibilities, heterodoxies, heresies and howlers remain. These are all my own work. Together (with whatever good bits), they make up my attempt to deal with the conundrum of organisms in evolution. If you read on you will be asked to entertain the idea that organisms are agents of a sort – not cognitive agents, but natural purposive systems nevertheless – and through this agency they enact evolution. Moreover, assimilating the agency of organisms into evolutionary thinking renders a conception of evolution that, while wholly consistent with Darwinism, puts considerable strains on the Modern Synthesis account of evolution with which we have all become so comfortably familiar. I can imagine that this message might not go down so smoothly with all readers. Never mind; I can always go back to painting squirrels.

Introducing organisms
Between unificationism and exceptionalism

Take a pinch (a gram) of good rich soil in the hollow of your hand; you may be holding half a billion organisms (Evans 2013). Add a drop of seawater (about 1 ml): you have probably enriched the ark in your palm by a million more (Becker 2013).[1] Not the most rigorous of surveys, perhaps, but any more systematic investigation would surely reinforce the same conclusion: the biosphere doesn't just *have* organisms, it is teeming with them. Organisms exist in a bewildering array of forms. Size alone hints at the scale of life's diversity; the mass of the largest organisms may exceed that of the smallest by over forty orders of magnitude.[2] The range of organisms' life activities is even more mind-boggling: they fly, they swim, they glow, they band together, they go it alone, they fix energy from sunlight, they live inside rocks and in deep-sea sulfur vents, they have contrived ways of surviving the most extreme and inhospitable places. Entire microbial ecosystems have just been discovered under the Antarctic ice (Christner et al. 2014). Moreover, these forms and activities exhibit a feature unique in the natural world; organisms are exquisitely suited to their conditions of existence. They are highly complex stable, adaptive, purposive systems. In the pursuit of their goals organisms possess a prodigious array of capacities. They are self-reproducing, self-building entities. They manufacture the very materials out of which they are constructed. These structures, these activities, this diversity, set organisms apart in the natural world. Organisms are natural entities to be sure, but they are no run-of-the-mill material things.

It isn't surprising, then, that organisms, their structures and activities, should demarcate a major scientific enterprise. Biology is the collection of disciplines dedicated to the study of living entities and their processes. There is hardly a

---

[1] You may also have killed a few.

[2] Blue whales (*Balaenoptera musculous*) may have a mass of 160,000 kg (NOAA Fisheries) while mycoplasma may be as little as 10^{-35} grams (Willmer, Stome and Johnston 2005). The General Sherman tree, *Sequioadendron giganteum*, in California is thought to have a mass of 560,000 kg (*Encyclopaedia Britannica* 2013). The fungus *Amillaria bulosa* may have a mass of 10,000 kg (Smith et al. 1992).

more vibrant intellectual endeavour in the world today. But modern biology embodies a conundrum.

The core of modern biology is the theory of evolution. It ranks among the most powerful, well-corroborated scientific theories ever devised. Its objective *inter alia* is to explain the fit, function and diversity of organisms (Lewontin 1978). Yet, the category *organism* has very little role to play in evolutionary theory. Nowadays, evolutionary theory principally deals in the dynamics of supra-organismal assemblages (populations) of suborganismal entities (genes). The distinctive properties that make organisms organisms play virtually no part in the explanation of evolutionary phenomena, or at least that version of evolutionary theory that has grown to such prominence throughout the twentieth century.

Perhaps this observation is too banal, too familiar, to raise any concern, but it is at least counterintuitive that the special properties that mark out a domain of scientific enquiry shouldn't have a place in the theoretical apparatus of that science. It would be decidedly odd, for instance, were the properties that make physical entities physical, or chemical entities chemical, or psychological entities psychological, not to figure in our theories of physics, chemistry and psychology. So, the absence of organisms from evolutionary theory ought to engender at least some mild curiosity.

It hasn't gone unnoticed, of course.

Something very interesting has happened to biology in recent years. Organisms have disappeared as the fundamental unit of life. In their place we now have genes, which have taken over all the basic properties that used to characterize living organisms. (Goodwin 1994:1)

But there isn't a whole lot of angsting about it either. Evolutionary biologists have tended to be rather sanguine about the incidental role of organisms: 'The organism is only DNA's way of making more DNA' (Wilson 1980: 3). Evolution, nowadays, is a molecular or genetical phenomenon.

Evolution is the external and visible manifestation of the survival of alternative replicators... Genes are replicators; organisms... are best not regarded as replicators; they are vehicles in which replicators travel about. Replicator selection is the process by which some replicators survive at the expense of others. (Dawkins 1976, 82)

Our objective here is to ask two questions about the place of organisms in evolutionary biology. The first is: 'Where did they go?' alternatively, 'Why have organisms been marginalised in evolutionary thinking?' The second is: 'Where should *we* go?' or 'in what ways, if at all, would our understanding of evolution be altered were organisms to figure more prominently?' In Part I, I survey some of the reasons for the marginalisation of organisms in twentieth-century evolution. In Part II, I canvass some reasons for reassessing this dis-organicised evolution. In Part III, I offer a rather programmatic, speculative attempt at a conception of evolution in which the distinctive capacities of organisms are pivotal. There is a prior question, of course; 'What *is it* about organisms that should have prompted this issue in the first place?'

I.1 Life's paradox

Dublin February 3, 1943. The Great Lecture Hall of Trinity College Dublin is buzzing with anticipation.[3] The first of three public lectures is about to be delivered by the newly appointed director of Trinity's School of Theoretical Physics. The speaker is the Nobel laureate physicist Erwin Schrödinger. Schrödinger's chosen topic may well strike the assembled audience as somewhat misplaced. He proposes to address the question *'What is life?'* Why should a physicist of such renown feel impelled to poach on biologists' preserve? As a pioneer of quantum mechanics, Schrödinger's illustrious reputation was forged through his imaginative engagement with some of the most perplexing problems of physics: wave mechanics, entanglement, nonlocality and that cat! Surely there was enough in the dynamic new physics to thrill and edify his expectant audience, without descending into the murky waters of biology. Most certainly there was, but it was the very dynamism of physics in the preceding half-century that underscored the urgency of Schrödinger's question. While the physics of Schrödinger's day had met with unprecedented success in revealing the fundamental laws that govern the world, those laws seemed to proscribe the very existence of living things.[4] How could life exist at all, let alone in its prodigious abundance, in contravention of the dictates of physics? What should the sciences do about organisms?

One strategy might be to adopt what we could call 'biological exceptionalism': biological phenomena are set apart as fundamentally different from the nonliving phenomena of the world. Organisms are unique, and we should acknowledge that their singularity earns them a special place in the sciences. There is no reason to suppose that they should be subject to mere physical law. Exceptionalism would certainly accommodate Schrödinger's observation that the laws of physics appear not to govern the behaviours of living things.

The idea of a law of nature seems to be a metaphorical extension of the juridical notion. Scientific laws, like their administrative counterparts, govern the behaviour of entities in their own proprietary domains. In Schrödinger's adopted institution, Trinity College Dublin, college law permits a student to order a glass of wine during an exam, but only if he is wearing his sword.[5] In

---

[3] The occasion is vividly depicted in Schneider and Sagan (2007).

[4] I'll say a little more about just how the laws of nature might have appeared to legislate against organisms below.

[5] It's always a 'he'.

Toronto, to cite another example, it is illegal to drag a dead horse down the street on a Sunday.[6] These rather parochial laws have no remit beyond their limited jurisdiction. Students at Trinity's rival institution, University College Dublin, are not similarly empowered to order wine during their exams. Nor does the stricture against hauling carcasses apply in, say, Calgary.[7]

Just as there are laws that apply only to Trinity and Toronto, there might be laws that hold only within specific scientific domains and not beyond. The laws of economics and psychology, if any such there be, evidently don't govern the behaviour of all living things, much less nonliving things. Perhaps the laws of physics no more govern behaviour in the biological realm than the by-laws of Toronto govern the behaviour of Calgarians. Biological exceptionalism suggests that the paradox of life that so perplexes Schrödinger may be a simple consequence of naively overgeneralising the laws of physics.

Exceptionalism of this sort would have been anathema to a physicist such as Schrödinger. Indeed, it would not at all sit easily with the vast majority of contemporary scientists and philosophers. The reason is that biological exceptionalism risks misrepresenting the relation between biology and physics. The respective domains of biology and physics may be distinct, but they are not disjoint. The laws of physics apply to all natural phenomena, including living things and biological process. The distinctive features and capacities of biological entities must be reconciled with the laws of physics.

We might call the alternative prompted by this line of thought 'unificationism' in contradistinction to exceptionalism. Unificationism, for its part, seems to honour a deep-lying modern commitment to the fundamental place of physics in the study of the natural world. But it has its downsides too. In particular, it fails to do justice to the very intuitions that motivate exceptionalism – viz. that biological phenomena are uniquely, and strangely different.

So, the dilemma raised by the paradox of life is how to understand organisms as, on the one hand, the highly aberrant distinctive entities they are, and on the other, mere physical things like anything else. Neither exceptionalism nor unificationism seems wholly satisfactory, yet each seems to capture a feature of the biological world that eludes the other. This is the paradox that Schrödinger's *What is Life?* addresses. As we shall see, the resolution he offers is highly original, enormously fertile and frustratingly ambiguous. But first it is worth noting that the dilemma Schrödinger addresses is as old as biology itself.

---

[6] Trinity: https://mligroup.wordpress.com/tag/trinity-college-dublin/; Toronto: http://www.lufa.ca/news/news_item.asp?NewsID=7235

[7] Indeed there's little else to do there on a Sunday.

I.2   Exceptionalism and unificationism

The problem of reconciling the naturalness of organisms with their peculiarity is an enduring challenge to the sciences. Indeed, an exceptionalist/unificiationist dialectic runs throughout the entire history of comparative biology.

I.2.1   Psuche

Aristotle, as Marjorie Grene (1974) was fond of saying, was the one great philosopher who was also a great biologist. Aristotle's extensive biological works set out to taxonomise and explain the startling array of biological forms, and their unique abilities to make their way in their respective conditions of existence. Aristotle is probably best considered an exceptionalist; his biological works emphasise the uniqueness of organisms. Each organism, Aristotle surmised, has a 'soul' (*psuche*). *Psuche* is nothing like the notion of soul as it appears in the Abrahamic religions; it is a theoretical posit, a wholly natural principle of organisation distinctive of organisms. In his principal work on the nature of life and mind, *de Anima*, Aristotle (1995) tells us that *Psuche* consists in an organism's goal-directed capacity to organise the matter at its disposal into a well-functioning organism typical of its kind.

Soul (*psuche*) is distinguished by a set of vital functions, nutrition, growth, locomotion and (in the case of humans) cognition. We may think of the form of an organism as a set of organizing principles, or a set of goal-directed dispositions, to organize its matter in such a way that the organism is capable of performing particular soul functions (in the particular way) distinctive of its kind. (Lennox 2001: 183)

The leading idea in Aristotle's biology is that the organism is a finely tuned functioning unit, whose parts and activities conjointly subserve the organism's '*way of life*'. The Aristotelian term is '*βιος*' (Lennox 2010). *βιος* consists in '... the full complement of an animal's activities organised around the single goal of its specific way of life' (Lennox 2009: 355). Each organism is imbued with a capacity to build and regulate itself in such a way that equips it to succeed in its distinctive manner. The world is replete with highly organised entities pursuing different, but related, ways of life.

Aristotle's biology is formulated in explicit opposition to that of the Pre-Socratic Atomists. The Pre-Socratic Atomists, like Democritus and Empedocles, are early exemplars of unificationism. The cosmos, according to the Atomists began in chaos, atoms moving randomly in the void. Some atoms chanced to encounter one another and combine. Some combinations were ephemeral, but some were stable and persisted. These aggregations of atoms give us our macroscopic, enduring entities, including organisms. The properties of any complex entity are simply the result of the causal, mechanical interactions of its parts. The entire edifice of the world is the consequence of chance encounters of randomly moving atoms, and the necessary consequences of their interactions.

Empedocles even posited an atomistic theory of the function and diversity of biological form that bears a passing resemblance to the modern theory of natural selection. Suborganismal parts were first formed by the agglomeration of earth. These disembodied parts – limbs, heads and the like – wandered the world aimlessly. Occasionally, under the guiding power of 'love' these parts aggregated together to form organisms. But, 'the power of love is a curious thing'; it makes for some mismatched partnerships. There were ox-headed men, and men-headed oxen. That all sounds arcane, and distinctly uncontemporary, until we ask why we no longer see such strange monstrosities. According to Empedocles, those collections of suborganismal parts that collaborated in making up well-functioning organisms survived; those that worked less well together perished.[8] For Pre-Socratic Atomists, then, organisms are mere congeries of their suborganismal parts. An organism's way of life is just the result of the mechanical interactions of its randomly aggregated pieces.

Aristotle insisted, against this picture of life as the consequence of random aggregations, that the chance encounters and mechanical interactions of an organism's parts are insufficient to explain the range and diversity of biological form. In fact, the Atomist attempt to explain an organism's way of life in terms of the arrangements of its parts inverts the proper explanatory order. For Aristotle, way of life explains the arrangement of parts and not the other way around. The co-ordination and integration of an organism's various parts is in no way haphazard. They are put there by the organism itself to subserve the organism's particular way of life. To suppose that these arrangements are merely matters of chance is to ignore some strikingly robust natural regularities.

What Aristotle means by this is that while any particular kind of organism may seem extremely improbable, it is also remarkably regular. Consider the extravagant 'sea slug' nudibranch *Pteraeolidia ianthina*.[9] These are bizarre, flambouyantly – almost psychedelically – coloured marine organisms that frequent the warm waters of the Indo West Pacific. They are a magnificent sight. Their iridescent blue cerata are striking, but they are not just for show. They house bioluminescent zooaxanthelae that fix energy from light, and provide nourishment for the nudibranch, enabling it to survive extended periods of time without eating – furbelow meets function. One might suspect that such a conspicuous creature would be easy prey. *P. ianthina* has a further

---

[8] Roux (2005) remarks: 'The account of the origin of species by Empedocles is the first recorded account of the theory of natural selection' (p. 6).

[9] The beast whose portrait graces the cover of this book.

contrivance to prevent itself from being eaten. It ingests the venomous spicules of sea anemones, which then pass through the slug and lodge in its mantle. The poison protects the nudibranchs from predators. This seems like a highly 'unlikely' – indeed outlandish – arrangement of form and function, and yet *P. ianthina* occurs reliably over and over again. The high fidelity reproduction of organisms capable of effectively pursuing the way of life typical of their kind, no matter how bizarre, is a robust regularity. Through their development and reproduction the same 'unlikely' forms occur over and over again, with the sort of predictability that we associate with truly law-governed phenomena, and not mere chance occurrences (Sorabji 1990). The regular occurrence of such highly integrated, exquisitely functional structures in organisms could not happen spontaneously (Johnson 2005). These regularities need to be explained, but to label them as mere chance, Aristotle believes, is to decline to do so.[10]

Aristotle asserts that some regularities in the world occur because they subserve purposes or goals. Systems that have goals or purposes reliably bring about states of affairs that are conducive to the fulfilment of those goals or purposes, however 'unlikely' they might have been. This is a very familiar way of predicting and explaining certain kinds of regular occurrences. Notably, we understand rational agency in this way. We explain and predict the behaviours of a rational agent by identifying her goals and then describing the ways in which her activities contribute to the fulfilment of those goals.

Where natural entities pursue goals or purposes, we can explain their behaviour and their structure by appeal to those purposes. Explanations that appeal to goals are called 'teleological'.[11] According to Aristotle, organisms are the very paradigm of purposive, goal-directed systems. The regular structures and activities of an organism, the intricate integration of its various parts, can all be explained by the fact that they are conducive to the pursuit of the organism's way of life. *Way of life* is a teleologically basic purpose of an organism, and it is what justifies teleological explanations and predictions of biological form and function in just the way that the goals of a rational agent underwrite the prediction and explanation of her actions. In representing organisms as entities that pursue purposes – ways of life – Aristotle is setting them apart from the rest of the natural world.

I.2.2   Modern mechanism

Aristotle's biology, ingenious as it might be, sounds a discordant note to the modern ear. The whole theoretical edifice has a thoroughly archaic cast. It

---

[10] I pursue the implications of this issue in more detail in Chapter 9.
[11] There are two recent superb extended discussion of Aristotle's teleology. They are Johnson (2005), and Leunissen (2010).

populates the natural world with forms, natures and purposes. Aristotelian science was anathema to those who forged the scientific revolution, from the arcane methods, to the esoteric metaphysics, to the biological exceptionalism. Indeed, it is often said that modern science was founded in explicit opposition to Aristotle. The scientists who initiated the Scientific Revolution – Copernicus, Galileo, Descartes, Gassendi, Boyle, Newton – sought to rid science of Aristotle's occult and obscurantist thinking.

The worldview that grew out of this revolution takes nature to be a machine.[12] We come to an understanding of the natural world in much the way that we learn about the workings of a machine. We take it apart (literally or conceptually) and investigate how the various parts fit together, and how the various interactions among the parts – pushing, pulling, bending, heating, repelling, attracting – aggregate to produce the activities of the complex whole. For half a century Descartes' physics provided the model for the study of the natural world, and the model of the natural world was the machine. 'I have described this earth, and indeed the whole universe, as if it were a machine: I have considered only the various shapes and movements of its parts' (Descartes 1647 [1985], §188).

Descartes' method of the machine applies just as much to his biology as to his physics. Organisms, for Descartes are *bêtes machines* – '*bêtes*' because, unlike humans, they are not possessed of a soul, a thinking substance; '*machines*' because their structure and function are entirely to be accounted for by the interactions of the parts.

The number and the orderly arrangement of the nerves, veins, bones and other parts of an animal do not show that nature is insufficient to form them, provided you suppose that in everything nature acts in accordance with the laws of mechanics. (1639 [1985],§134)

Living or otherwise, machines are machines.

Descartes' mechanistic biology is much praised by his contemporaries, and highly influential. Philip Sloan (1977) cites the sixteenth-century Danish ana-tomist, Neil Stenson's, effusive endorsement of the power of Descartes' mechanism to reveal the mysteries of even human anatomy:

Descartes . . . was the first who dared to explain all the functions of man, and especially of the brain, in a mechanical manner. Other authors describe man; Descartes puts before us merely a machine, but by means of this he very clearly exposed the ignorance of others who have treated of man, and opened up for us a way by which to investigate the use of the other parts of the body as no one has done before. (Quoted in Sloan 1977: 20)

---

[12] The methodology of mechanism is treated in more detail in Chapter 1.

Thus Descartes' conception of living things embodies a radical unificationism, set explicitly in opposition to the exceptionalism of the ancient and medieval study of organisms.[13]

I.2.3   Organisms as natural purposes

Despite Descartes' advocacy, many early modern biologists and philosophers were distinctly pessimistic about the capacity of the new science to subsume all the distinctive features of the biological world. One enduring problem for biologists was how to explain the development of an organism from its comparatively modest, undifferentiated single-celled zygote stage to the fully formed, complex highly differentiated, integrated adult. To many eighteenth-century biologists, like Georges Comte de Buffon and Caspar Friedrich Wolff (*contra* Descartes' insistence), there seemed to be too little specific detail contained in a fertilised egg to allow one to 'deduce from that alone, . . ., the whole figure and conformation of its parts' (Wolff: quoted in McLaughlin 2014: 8). These biologists surmised that there must, then, be some extra 'epigenetic' factor, not contained within the fertilised egg, that directed the organism toward the development of its proper form. For some this 'penetrating force' or '*vis essentialis*' is a wholly biological, nonphysical feature of the world.

The philosopher Immanuel Kant's biology was strongly influenced by Buffon and Wolff (among others). He absorbed from Wolff the evident inability of mechanism to explain regularities of organismal development. He took from Buffon the idea that organisms are self-organising, self-synthesising entities. Yet, as a thoroughly modern scientific thinker, Kant was keenly aware of, and strongly supportive of, the modern, mechanical conception of science. He considered Newton's laws of motion to have limned the very nature of matter. Newton's laws tell us that matter is inert, non-self-moving. When it changes, it does so through the influence of external forces acting from outwith. Matter does not organise itself; it does not replicate itself. But organisms do all these things. If Newton's laws lay down the rules by which matter conducts itself, organisms flout them flagrantly. Kant expresses his pessimism that the methodology that revealed the secrets of physics should be so forthcoming with those of biology (Cornell 1983).

Kant makes much of the fact that organisms synthesise the materials out of which they are made (McLaughlin 2000). Consequently, an organism has the parts it has, in their particular arrangements, precisely because in its

---

[13] The machine analogy persists into contemporary biology (see, for example, Lewens 2004), where Nicholson complains, it '. . . fundamentally misrepresents the nature of the very subjects . . . it seeks to explain' (2014: 168–169). Talbot (2013) argues for the importance of the 'machine-organism' concept in generating the Modern Synthesis conception of the unit gene.

development the organism has made those parts and has put them there to serve its crucial vital functions:

> ... an organism first processes the matter that it adds to itself into a specifically-distinct quality, which the mechanism of nature outside of it cannot provide, and develops itself further by means of a material which, in its composition, is its own product. (Kant 1790 [2000]: 371)

As a consequence, the relation between an organism and its parts is vastly different from the relation between a run-of-the mill machine and its parts. Like machines, organisms are the consequences of the interactions of the parts. But, unlike machines, an organism's parts are the consequence of the activities of the organism as a whole. The constituent parts and processes of a living thing are thus related to the organism as a whole by a kind of 'reciprocal causation'.

> I would provisionally say that a thing exists as a natural end if it is cause and effect of itself. (Kant 1790 [2000]: 371)

Organisms are thus crucially unlike machines, just as Buffon insists. They are self-building, self-regulating, highly integrated, functioning wholes, that exert a particularly distinctive influence over the capacities and activities of their parts.

> The essential definition Kant offered of ... organic form ... was that of the reciprocal interrelation of parts as means to ends, and consequently of the priority of the whole over the parts as means and ends, and consequently, of the priority of the whole over the parts in the constitution of the entity. (Zammito 1991: 218)

Kant's own definition of an organism emphasises the reciprocity between part and whole.

> The definition of an organic body is that it is a body, every part of which is there for the sake of the other (reciprocally as an end, and at the same time, means).... An organic (articulated) body is one in which each part, with its moving force, necessarily relates to the whole (to each part in its composition).[14]

Kant is simply highlighting the distinguishing property of organisms that we have already identified: their purposiveness. He tells us that purposiveness is essential to our concept of an organism. Moreover, an organism's purposes are immanent in it; they are not extrinsic purposes, like those of artefacts. Hence, organisms are natural purposes.

Their status as natural purposes puts organisms beyond the ambit of the methods of the modern sciences. The features and capacities of organisms just cannot be adequately accounted for by appeal to causes and laws (Ginsborg 2001). Thus while endorsing the modern conception of the scientific enterprise,

---

[14] *Opus postumum*, quoted in Guyer (2005: 104).

Kant acknowledges a gap between the phenomena that science can bring under its ken, and the phenomena that mark out the biological world as a distinct domain. And he despairs of ever closing it:

> ... it is quite certain that we can never adequately come to know the organised beings and their internal possibility in accordance with merely mechanical principles of nature, let alone explain them; ... it would be absurd for humans even to make such an attempt or to hope that there may yet arise a Newton who could make comprehensible even the generation of the a blade of grass according to the natural laws .... (1790 [2000]: 400)

Kant is endorsing a robust version of biological exceptionalism here, in explicit opposition to the unificationist predilections of modern mechanism.

I.2.4   Vitalism

Kant's pessimism about biological unificationism is clearly influenced by the vitalism of certain seventeenth-century biologists. 'Vitalism' is an umbrella term for any of a number of variously radical forms of biological exceptionalism.[15] In its simplest version, vitalism can be seen as a metaphysical thesis. It holds that if all explanations in science appeal to forces, and the material forces of the mechanistic philosophy are inadequate to explain biological form, then there must be other, nonmaterial forces that cause organisms to be functionally integrated, and exquisitely organised as they are.[16] Vitalism had some currency among developmental biologists into the eighteenth century and latter part of the nineteenth, lingering even into the early twentieth century.

A vivid example is to be found in the German developmental biologist Hans Driesch. His vitalism was inspired by the results of some remarkable experiments on sea urchin embryos conducted in 1891 (Allen 2005). One question puzzling Driesch, a version of the same problem that perplexed the seventeenth-century epigeneticists, like Caspar Friedrich Wolff, was how an organism's tissues and organs can differentiate from undifferentiated precursor cells.[17] The prevailing model at the time had been proposed by Willhem Roux and August Weismann. It suggested that at each cell division each of the daughter cells received only a portion of the hereditary material. That portion directed the cell to develop into its particularly specialised form. Driesch's experiments on sea urchin embryos sought to confirm this hypothesis. He divided embryos in half at the 16-cell stage. To his surprise, each half embryo developed into a complete sea urchin larva. Driesch reasoned that each

---

[15] I am greatly indebted to Mike Stuart for an enlightening tutorial on vitalism.

[16] There are also, however, methodological versions of vitalism that do not make these metaphysical pronouncements (Wolfe 2011).

[17] It is still one of the crucial questions in developmental biology. As C.H. Waddington asked: 'Exactly, how does an egg produce legs, ... ?' (1966: iv). (I encountered this quotation initially in Sansom 2011: 23.)

partial embryo possessed the resources, and the material causes, to produce only *half* of a larva. Yet each was able to compensate in such a way as to produce an entire organism. No mere machine could build itself out of the material sufficient for making only some of its parts. There must then be some further force or cause that guides the development of a complete embryo out of its parts.

He claimed that embryonic development was guided by an entelechy, an organizing, directive force that consumed no energy, was immaterial, but was the factor that distinguished living from non-living matter. (Allen 2005: 271)

Driesch posited a *vis essentialis*, a nonmaterial principle that guided organisms toward the attainment of their complete, proper form.

Nonmaterialist vitalism rapidly fell from favour for a number of reasons. Principal among them was that it was so comprehensively out of keeping with the modern scientific conception of the world. An argument from the philosopher Moritz Schlick (1953) serves to establish how deeply incompatible substance vitalism is with the commitments of modern physics. Schlick pointed out that according to the presuppositions of modern physics, the material world is causally closed. That means that any cause of an event realised in matter, must also be realised in matter. Consequently, there can be no nonmaterial causes, as vitalists take their *vis essentialis* to be. Schlick's argument, of course, doesn't refute vitalism, but it does highlight the way that its sheer exceptionalism mounts a *prima facie* case against it.

Vitalism is now often ridiculed as vapid and vacuous:

To account for the phenomena of life by ascribing them to *vitalism* is no more helpful or intelligible than to explain the properties of water as due to *hydrism* or of light to *photism*. (Conklin 1921: 354)

So thoroughly has vitalism been rejected since the early twentieth century, that it and its cognates have become terms of abuse:

... and so to those of you who may be vitalists I would make this prophecy: what everyone believed yesterday, and you believe today, only cranks will believe tomorrow. (Crick 1967: 99)

As a form of exceptionalism, vitalism lost out to the unificationist inclinations of twentieth-century science. Indeed, any lingering appeal of vitalism appears to have been expunged by the staggering success of the molecular revolution in biology of the twentieth century, which pursued an unabashedly unificationist programme: life as chemistry (Morange 2008).[18]

---

[18] This issue is taken up in much greater detail in Chapter 1.

I.2.5   Against mechanism

Exceptionalism doesn't entail substance vitalism, but it may require the rejection of modern mechanism. A tradition of materialist anti-mechanism runs through in the fringes of much of twentieth-century biology, even among those who considered themselves materialists in good standing. The great American botanist, Agnes Arber writes:

> The *mechanist*, starting from the physico-chemical standpoint, interprets the living thing by analogy with a machine. The *vitalist*, on the other hand, supposes a guiding entelechy, which summons order out of chaos; he thus adopts a dualistic attitude. The elements of truth in both these views are recognised, and their opposition is resolved in the *organismal* approach to the living creature. This approach is conditioned by the belief that the vital co-ordination of structures and processes is not due to an alien entelechy, but is an integral part of the living system itself. (Arber 1964: 100–101)[19]

Likewise, E.S. Russell's plea for the distinctiveness of organisms calls for a rejection of mechanism and the recognition, reminiscent of Aristotle, that organisms alone are purposive entities. Russell is under no illusions about the profound implications of his views, as the preface of his *The Directiveness of Organic Activities* makes plain.

> My rejection of mechanism is quite deliberate and for a good cause. The living thing can be treated as a physico-chemical system or mechanism of great complexity, and no one would dream of denying the validity and value of biochemical and biophysical research. But such an approach leaves out of account all that is distinctive of life, the directiveness, orderliness and creativeness of organic activities, and completely disregards its psychological aspect. I try to show that we cannot disregard these unique characteristics of life without losing all hope of building up a unified, coherent and independent biology. (Russell 1945: viii)

So, this much, at least, appears evident, that even the most cursory trawl through two and a half millennia of biology serves to underscore the recurring and persistent challenge posed by the nature of organisms to the natural sciences. Understanding organisms places competing, some have thought irreconcilable, demands on the natural sciences. Organisms are at once part of the unified structure of the natural world, and wholly exceptional. It is hardly surprising, then, that the question that Schrödinger posed to the expectant crowd in Dublin on that Friday night in 1943 should seem so, well, *vital.*

I.3   Schrödinger's *'What is Life?'*

In fact Schrödinger's question is so good that he answers it twice. His two responses are very different in character, and it is fair to say that they haven't

---

[19] I thank Dick Vane-Wright for directing me to this passage. See his (2014) for a discussion.

fared equally well. It isn't entirely clear that Schrödinger was aware how different his two responses were, or how wildly divergent their prospects were to be.[20]

Schrödinger begins by making a point with which by now we are well familiar. Life appears to violate the laws of nature. Whereas, Kant considered living things to violate Newtonian laws, the laws that concern Schrödinger are those of thermodynamics, in particular the second law. The second law of thermodynamics tells us that a closed system, one in which energy neither enters nor leaves, will spontaneously move toward a state of maximum entropy and stay there. That is to say it will move toward an arrangement in which energy is evenly distributed throughout, a state of maximum *disorder*. Order decreases, as a matter of law. But organisms *increase order* by building highly structured, stable forms, in apparent violation of the second law. Schrödinger called organisms 'negentropic systems'. How can organisms build, maintain and replicate such astonishing order in the face of the universal tendency of the world to generate *disorder*? Schrödinger's question '*What is Life?*' thus boils down to asking how organisms escape the grip of the second law. His two answers suggest two quite different strategies.

Order from order   Some naturally occurring entities resist thermo-dynamic decay by being tightly structured and highly stable. Crystals are the best examples. A diamond, for instance, is constituted of an aggregate of carbon atoms arranged in a characteristic tetrahedral structure. Each Carbon atom is joined to four others forming a face-centred cubic lattice. This structure famously confers on diamonds their notorious hardness, and durability. Naturally occurring diamonds may be up to 3.3 billion years old, and have maintained their integrity despite being transmitted to the surface of the earth immersed in molten magma. Schrödinger supposes that one way in which organisms might escape the threat of thermodynamic decay would be to be built and maintained by a stable, robust crystal.[21]

But there is a problem with this crystal hypothesis. While the crystalline structure imparts stability to a diamond, it severely constrains the variety of forms that a diamond can take. A diamond simply repeats the same structure over and over; it is a periodic crystal. Given the limited range of forms that a periodic crystal can take, it seems implausible that the staggering diversity of life forms could be underwritten by the periodic structure of crystals. Schrödinger thus hypothesises that the properties of living things must be secured by a crystal that has an 'aperiodic' structure. An aperiodic crystal is

---

[20] Kauffman (2010) offers a characteristically engaging discussion of the various implications of Schrödinger's two answers.

[21] He credits the physicist Max Delbrück with this insight.

one whose structure is open-ended. As a crystal this structure has a great deal of stability, as 'aperiodic' it has the capacity to become indefinitely complex without repeating itself.

The aperiodic nature of these hypothetical crystals would allow them to act as a 'code-script' encoding information about the construction of like organisms. These molecules

play a dominating role in the very orderly and lawful events within a living organism. They have control of the observable large scale features which the organism acquires in the course of its development, they determine important characteristics of its functioning; and in all this very sharp and very strict biological laws are displayed. (Schrödinger 1944: 20)

The order-from-order solution appears to have inspired much of the molecular biology revolution. The rise of molecular biology in the twentieth century was fueled by the conviction that all the properties of life could be explained as manifestations of the properties of complex molecules. The discovery of the structure of DNA by Watson and Crick ten years after Schrödinger's lectures appears to have fulfilled his prophecy. DNA is a crystalline structure, of sorts, but one capable of unending variety. It is also capable of making itself into a template for the copying of other, identically structured crystals. The structure of DNA revealed to Watson and Crick the way that information about building an organism could be encoded in a crystal, copied and passed on unchanged from one generation to another. It appears to have precisely the properties that Schrödinger surmised would be necessary to underwrite the definitive feature of life, its ability to resist thermodynamic decay while taking on an unrestricted variety of forms.

In the paper announcing their discovery Watson and Crick comment, somewhat unostentatiously: 'This structure has novel features which are of considerable biological interest' (Watson and Crick 1953). James Watson was evidently less circumspect out of print. Legend has it that he walked into the Eagle Pub in Cambridge on 28 February, 1953, announcing that he had discovered the 'secret of life'. He takes up the theme in the first edition of his *Molecular Biology of the Gene*.

We see not only that the laws of chemistry are sufficient for understanding protein structure, but also that they are consistent with all known hereditary phenomena. Complete certainty now exists among essentially all biochemists that the other characteristics of living organisms ... will all be completely understood in terms of the coordinative interactions of small and large molecules. (Watson 1965: 67)

Schrödinger's 'order from order' answer suggests a strong version of unificationism. The key to understanding life is to understand the dynamics of these highly structured molecules *as molecules*.

So successful has that project been that the very concept of life has become molecularised (Morange 2009).

... the development of the implications of the Watson-Crick model have given us persuasive reasons for believing that biology is nothing more than chemistry – but chemistry, nonetheless, in which the chemical systematization of chemical elements plays a most important role. (Schaffner 1969: 326–327)

Jacques Monod endorses this molecular turn.

In a sense, a sense very real, it is at this level of chemical organization that the secret of life lies, if indeed there is any one secret. And if one were able not only to describe these sequences but to pronounce the law by which they assemble, one could declare the secret penetrated, the *ultima ratio* discovered. (Monod 1971: 96–97)

Indeed, so comprehensive has this unificationist programme been that much of later twentieth-century biology has operated under the assumption that an account of the nature of living things is unnecessary for biology; life is chemistry.

Order from disorder   Schrödinger's second answer, however, takes a significantly different tack. He pointed out that organisms exist far from thermodynamic equilibrium. They occupy a milieu in which there are enormous gradients in the distribution of energy. Moreover, they are open systems, systems in which matter and energy are constantly being assimilated and emitted. The second law predicts that open systems in steep energy gradients like this will dissipate energy rapidly. Sometimes the most efficient way to dissipate energy is to build structures that do so. Organisms do just this. In creating these energy-dissipating structures they increase the disorder around them.

So, on this second response, organisms are dissipative systems. They avoid thermodynamic decay by actively exchanging energy and matter with their environments. They take in energy in the form of food, sunlight, and they disperse it as heat and waste. This, Schrödinger says, is the key for understanding biological order. Organisms are highly negentropic, it is true, but they do not violate the laws of thermodynamics. An organism 'pays its debt to the second law' by increasing the entropy around it. Metabolism is the process whereby organisms build order internally, while decreasing it in their environs:

... by eating, drinking and breathing and (in the case of plants) assimilating. The technical term is *metabolism*. The Greek word ... means change or exchange. Exchange of what? ... What then is that precious something contained in our food which keeps us from death? That is easily answered. Every process ... everything that is going on in Nature means an increase of the entropy of the part of the world where it is going on. Thus a living organism continually increases its entropy – or, as you may say, produces positive entropy – and thus tends to approach the dangerous state of maximum entropy, which is death. It can only keep aloof from it, i.e. alive, by continually drawing from its environment negative entropy – which is something very positive as we shall immediately see ... Or to put it less paradoxically, the essential thing in metabolism is that the organism succeeds in freeing itself from all the entropy it cannot help producing while alive. (Schrödinger 1944: 71–72)

Organisms are not unique in their capacity to build spontaneous order, except in the degree to which they do so. If we heat a layer of oil from beneath, we observe that the column of oil forms a highly ordered, complex arrangement of convective cells (Bénard cells). This is a stable arrangement of tightly packed hexagonal cells all of the same size. Each cell consists of a central column that conveys hot liquid from the lower surface to the top, and a descending periphery that conveys the cooled liquid back to the bottom. This arrangement is stable; if it is perturbed, the system will spontaneously return to the optimal arrangement of tightly packed cells of the same size and shape. Bénard cells exhibit an incipient form of the kind of self-building, self-regulating behaviour that is so distinctive of organisms.

Given the unificationist/exceptionalist dialectic running through the history of biology, Schrödinger's 'order from disorder' response suggests something new, a kind of *détente*, or synthesis. Far from equilibrium systems, are consistent with, even predicted by, the laws of physics. Yet they are qualitatively different from near equilibrium systems.

In physics, the theory of open systems leads to fundamentally new principles. It is indeed the more general theory, the restriction of kinetics and thermodynamics to closed systems concerning only a rather special case. In biology, it first of all accounts for many characteristics of living systems that have appeared to be in contradiction to the laws of physics, and have been considered hitherto as vitalistic features. Second, the consideration of organisms as open systems yields quantitative laws of important biological phenomena. (von Bertalanffy 1950: 23)

Organisms are like nothing else in the natural world, but not despite the laws of physics, because of them.

In order to understand these systems, we should look to their dynamics. In fact in far from equilibrium systems it would appear necessary to study the dynamics of the entire system in order to explain the behaviour of the parts. We understand the movements of a water molecule in a convective cell, for instance, by understanding that it is part of the dynamics of the system as a whole. Schrödinger seems well aware that his order from order response calls for a new kind of thinking about laws and dynamics.

[F]rom all we have learnt about the structure of living matter, we must be prepared to find it working in a manner that cannot be reduced to the ordinary laws of physics. And that not on the ground that there is any 'new force' or what not, directing the behaviour of the single atoms within a living organism, but because the construction is different from anything we have yet tested in the physical laboratory. (Schrödinger 1944: 76)

The 'order from disorder' approach has received much less attention from mainstream evolutionary biologists than the order from order alternative. It has not figured prominently in the growth and development of twentieth-century evolutionary theory. Nevertheless, it too has provoked some interest. The dynamics of far-from-equilibrium systems has been the subject of a lively research programme in physics (Prigogine and Stengers 1984). It has inspired the science of complex systems dynamics, which has investigated the principles of self-organising, self-regulating systems. The capacity to build structures that efficiently dissipate energy, to adapt and maintain stability are spontaneous features of matter far from thermodynamic equilibrium. The order of the biological world comes for free:

[M]ost of the beautiful order seen in ontogeny is spontaneous, a natural expression of the stunning self-organization that abounds in very complex regulatory networks. We appear to have been profoundly wrong. Order, vast and generative, arises naturally .... (Kauffman 1993: 25)

Schrödinger's second response to the paradox of life suggests that in their growth development and organisation, organisms are extreme examples of what happens to a system far from thermodynamics equilibrium. The order in organisms is natural, and to be expected. In order to explain these features of organisms, one should then look to the capacities of entire organisms. If these, in turn, are relevant for understanding the process of evolution, then a suggestion imposes itself; organismal dynamics should be an integral part of the theory of evolution.

That there are two ways of understanding life – one based on the underlying molecular mechanisms of organisms, the other on their gross dynamics – pinpoints the dialectical tension of this book. What is the relation between these approaches? Why has Modern Synthesis evolutionary theory chosen to prioritise the molecular mechanism approach at the expense of the other? Is there a viable evolutionary theory that places the dynamics of organisms in the foreground?

Conclusion

Organisms are wholly natural and entirely singular features of the world. Modern evolutionary biology has dealt with the strangeness of organisms by the simple expedient of ignoring them. The purposive complexity of organisms is seen as simply a consequence of evolution, which, in turn, according to currently orthodox Modern Synthesis thinking is basically a molecular process. The wonderful order of the biological world is built and sustained by genes.

Yet, we have just sampled what looks to be a tantalising alternative. The magnificent complexity and functional integration of organisms is something of their own making. It arises out of their intrinsic dynamics as far from thermodynamics equilibrium systems. These are both wholly natural, scientifically reputable approaches to understanding the nature of organisms.

It is perhaps understandable that if evolution is primarily a phenomenon of genes, that the dynamics of organisms might be considered of only peripheral interest. On the other hand, if evolution is mediated and structured by the remarkable properties of organisms, then it might seem that little is to be gained, and potentially much to be lost, by leaving them out. This is just what the Modern Synthesis theory of evolution has chosen to do.

One pressing question, then, is why? Why did Modern Synthesis evolutionary biology choose to pursue a strategy that concentrates upon the capacities of molecules, rather than one that emphasises the distinctive dynamics of organisms? Why did twentieth-century evolutionary thinking pursue the 'order from order' agenda, at the evident expense of the 'order from disorder' approach? Why is evolution a phenomenon of genes rather than organisms. The story of the extirpation of organisms from evolution, at the expense of genes, is a complicated one. We take it up in Part I, before exploring the prospects of an organism-centred evolution.

Part I

The eclipse of the organism

In periods of what Thomas Kuhn (1970) called 'normal science', the disciplinary matrix plays a distinctive role. 'Disciplinary matrix' is Kuhn's term for the set of norms, rules, metaphysical beliefs, methods and theoretical concepts that serve as a sort of template for scientific practice at a given time. Scientists are initiated into it in their training; they draw upon it to define their research programmes, to determine what counts as a reasonable, answerable question, and what counts as an acceptable answer. The conduct of normal science depends upon the compliance of the members of a discipline with the terms of the disciplinary matrix. At such times, the relation between scientific research and the disciplinary matrix that motivates it only really goes one way. 'Kuhn is keen to emphasise that normal science, whether experimental or theoretical, is low on innovation and high on dogma . . . in normal science, the theory is not up for debate. Research takes it for granted; it therefore has the status of dogma' (Bird 2000: 36–37).

If there have been any genuine episodes of normal science, the evolutionary biology of the mid-twentieth century onward must surely be one. Modern Synthesis evolutionary biology is largely the consequence of the resolute pursuit of the order-from-order conception of living things that we encountered in the Introduction. It has given us a special kind of molecular entity – the gene – as the canonical unit of biological organisation. Genes are theoretically privileged things: they are discrete units of phenotypic control, exclusive units of inheritance and units of evolution. Genes severally encode instructions for building phenotypes; together they amount to a blueprint for constructing an entire organism. These instructions are copied and passed from parents to offspring in inheritance. Moreover, change in the structure of a population counts as evolutionary change only if it is realised as change in the relative frequency of gene types. One consequence of prioritising genes in this way is that it fosters the impressions that what happens within an organism, downstream of conception, has little if any direct influence on evolution. That goes for behaviour, development, social interaction and learning; innovations brought about by these processes are not evolutionary innovations.

No one can gainsay the successes of the Modern Synthesis theory of evolution, particularly its reliance on the gene as its principal theoretical unit. But at some point, in any disciplinary matrix, it behooves those under its sway to hold the conceptual underpinnings of their theory up to heightened scrutiny. Sometimes the impetus to do so is generated from empirical work done under the auspices of the disciplinary matrix.

Such an opportunity was provided, for example, by the Human Genome Project. At its inception, the director of the project, Harry Collins, announced that the objective of the programme was 'to characterize in ultimate detail the complete set of genetic instructions of the human being' (Collins 1999: 28). It was estimated that the human genome should contain 80,000–100,000 genes (Keller 2011a). The results have been puzzling, and to some disappointing. The current estimate of human genes now stands at a comparatively paltry figure, somewhere between 20,000 and 23,000, scarcely any more than the 19,000 or so genes in the decidedly less complex *C. elegans* (Collins 1999). It turns out, surprisingly, that only 1.5 per cent of the human genome codes for proteins that build organismal structures. Yet at the project's outset this had been taken to be the prime function of genes. Nor, as it transpires, is the noncoding DNA 'junk', as it was previously thought to have been (Keller forthcoming). Increasingly it is becoming obvious that this DNA too plays a crucial, highly complex role in genome function.

All in all, it seems the disciplinary matrix in which the Human Genome Project was embedded has generated empirical results that diverge quite radically from those it predicted. This provides the opportunity to reassess the concept of the gene. At the same time, it invites us to revisit the theory of evolution that the concept of the gene did so much to inspire.

Other considerations suggest that a re-evaluation of the Modern Synthesis conception of evolution is in order (Laland et al. 2014; Wray et al. 2014). One particularly compelling one is the advent of what is being called 'evolutionary-developmental biology' – 'evo-devo' (Carroll, Grenier and Weatherbee 2000; Hall 1999; Irschick, Albertson and Brennan 2013; Morange 2011) – and its complementary successor developmental-evolutionary biology – 'devo-evo' (Laubichler 2009; Maienschein and Laubichler 2014). They share a common ground; each presses the claim that the heretofore neglected processes of organismal development are crucial to an understanding of evolution.

A good place to start this re-evaluation might be to ask how we arrived at our gene-centred Modern Synthesis theory of evolution. At the same time, it will be salutary to ask how as evolutionary biology promoted the gene as its unifying theoretical entity, it came to marginalise the organism. Why, as the order-from-order approach to understanding life was comprehensively adopted by twentieth-century evolutionary theory, was the order-from-disorder approach neglected? The reasons are complex and intertwined; they are historical, conceptual and empirical. The objective of this section is to isolate and diagnose some of them. It comprises three chapters, each of which amounts to a specific conjecture about the rise of gene-centred evolutionary biology at the expense of an organism-based alternative.

Chapter 1 makes the case that a significant part of the reason is methodological. Modern evolutionary biology has been guided by a set of methodological precepts about the proper conduct of science. The precepts fall under the loose rubric 'mechanism', and they have been with us, driving the enormous success of modern science, since the advent of the Scientific Revolution. The study of life as a phenomenon of order arising from molecular order is cozily congenial to mechanism. While this mechanism-inspired biology has been extremely fecund, I want to suggest that it may not be the only game in town. There is the prospect of an order-from-disorder evolutionary biology that has been consistently overlooked and undervalued. One obvious reason for the neglect is that such an approach doesn't fall so conveniently within the ambit of mechanism. Chapter 2 surveys a compatible, yet alternative explanation for the alienation of individual organisms. Evolutionary biology is largely a theory about population dynamics. As it has developed from Darwin's early formulation to the highly sophisticated version enshrined in the Modern Synthesis theory, it has taken on a new ontology and a new theoretical apparatus. Darwinian populations are composed of organisms; those of the Modern Synthesis theory comprise abstract trait (gene) types. The transition from Darwin's theory to its Modern Synthesis successor is one in which the theoretical role of individual organisms has been completely displaced by that of abstract trait types. Modern Synthesis population dynamics is an enormously more powerful, quantitative approach to evolution. Here again, though, it is worth pausing to weigh the costs. We ask whether Modern Synthesis population dynamics captures all that can be explained by Darwin's own less rarefied theory. Chapter 3 presents yet another reason why organisms may have lost their place. One of the great theoretical dividends of gene centrism has been a simplification and streamlining of our concept of evolution. Evolution, generally construed, comprises four component processes: inheritance, development, adaptive change and novelty. Genes play a unique role in each component process. The great advantage of gene-centrism is that it has allowed us to cast these processes as discrete and independent. It may well be, however, that the advantages of convenience are outweighed by the disadvantages of fractionation. In particular, it is beginning to be realised that the fractionated picture distorts the relation between the component processes of evolution. As we see in Part Two, this is nowhere more pronounced than in the marginalisation of organismal development ushered in by the Modern Synthesis.

1   Mechanism, reduction and emergence
Of molecules and method

That laughter emanating from the salons of seventeenth-century Paris is the sound of the passing of a philosophical tradition. The source of the mirth, John-Baptise Poquelin (Molière), has just introduced the assembled Parisian intellectuals to his pathetic, effete aristocrat *Le Malade Imaginaire*, whose doctor explains that the medication he has administered induces sleep by dint of its *virtuus dormitiva* (dormitive virtue).[1] Of course the assembled *haut monde* know full well that the object of Molière's rapier wit isn't hypochondria, nor is it so much the mores of the ruling elite. It is a mode of thought – Scholasticism – that had dominated Europe for half a millenium.

On the caricature that Molière is peddling, the science of the Scholastics is fatuous and feeble. Their world is suffused with arcane substantival forms and final causes, and they content themselves with vacuous platitudes masquerading as understanding, to the extent of accepting that the substantival form of a sleeping draft – its dormitive virtue – explains its capacity to put people to sleep. In deriding the old science, Molière is propagandising for the new. His biographer and fellow satirist, Mikhail Bulgakov, credits the French philosopher and astronomer Gassendi with turning Molière's head from the science of the schools to modernity: 'Gassendi instilled in him a love for clear and precise thought and hatred of scholasticism, respect for empirical experience and contempt for falseness and flowery bombast' (Bulgakov 1986: 35). For his own part, Molière grasped that what Scholasticism may have lacked in intellectual rigour, it more than made up for in comic potential.

The bold new science, against which Molière uses Scholasticism as a foil, has come to be known as 'mechanism'. 'Mechanism' is a sort of umbrella term intended to gather a collection of views about the appropriate way to do science emerging around this time. To be sure, there is (and was) some debate about who was and wasn't a genuine mechanist, and what precisely 'mechanism'

---

[1] Ever the master of irony, Molière died of consumption in the middle of a performance of this play, erroneously believing himself to be well enough to play the hypochondriac erroneously believing himself to be mortally ill.

stood for. There is little disagreement about what mechanism stood against, or about its significance. As Daniel Garber notes, 'mechanism'

has become kind of a catch-all category that is taken to include most of those who opposed the Aristotelian philosophy of the schools, ... part of a broad master narrative about the demise of scholastic Aristotelian philosophy ... and the rise of modern mathematical and experimental science. The titanic clash that gave birth to modernity. (Garber 2013: 3)

More positively, mechanism seems to have involved a comparatively minimalist view of the make-up of the world, and a corresponding humility about how best to study it. For Molière's teacher, Gassendi, and for Boyle, for example, mechanism consisted in the rather radical idea that the natural world is simply and exclusively matter in motion. The job of science is to find out what makes matter move. Accordingly, it should restrict itself to the observation, measurement and manipulation of matter and motion.

As is often the case in science, the growth of mechanism seems to have crystalised around a metaphor rather than an explicit set of rules and principles.[2] In this case, the metaphor is the clock.

Of all the mechanical constructions whose characteristics might serve as a model for the natural world, it was the clock more than any other that appealed to many early modern natural philosophers. Indeed, to follow the clock metaphor for nature through the culture of early modern Europe is to trace the main contours of the mechanical philosophy, and therefore of much of what has been traditionally construed as central to the Scientific Revolution. (Shapin 1996: 32)[3]

The task of the scientist is to enquire into the mechanics of the clockwork world. A clock may be designed, as a whole, to pursue a particular function, but it is built out of its parts. The best way to understand its workings is to take it apart, and consider 'the various shapes and movements of the parts'.

On the emerging mechanist worldview, the edifice of science is to be built by procedures guaranteed to deliver certain knowledge. The methods of science must eschew the reliance upon occult entities such as insubstantial substances, essences or inner strivings for perfection, and must restrict itself to only those features of the world that can be measured, manipulated and deduced from clear and certain intuitions. Only these can yield scientific certainty, and the only features of the world that those procedures can reveal are the motions of, and the causal relations between, material entities. René Descartes is among those credited with forging this new conception of scientific methodology.[4] In his *Rules for the Direction of the Mind*, he enjoins that

---

[2] 'Metaphor and simile are the characteristic tropes of scientific thought, not formal validity of argument' (Harré 1986: 7).
[3] I thank Cory Lewis for bringing this passage to my attention.
[4] Not that this endeared him particularly to either Molière or Gassendi.

the proper conduct of science proceeds by analysing complex phenomena
into their simple constituents.

We shall be following this method exactly if we first reduce complicated and obscure
propositions step by step to simpler ones, and then, starting with the intuition of the
simplest ones of all, try to ascend through the same steps to knowledge of all the rest.
(Garber 1998/2003)

The injunction to reduce the complex to the simple applies just as much to the
understanding of physical systems as it does to 'propositions'. The investigation of a complex physical system proceeds by breaking it down into its
component parts and observing how those parts fit together, and how their
individual activities contribute to the functioning of the system as a whole.[5]

Nancy Cartwright calls the method inaugurated by the Scientific Revolution,
the 'analytic method'. It is still the principal mode of scientific investigation
today:

[T]o understand what happens in the world, we take things apart into their fundamental
pieces; to control a situation we reassemble the pieces, we reorder them so they will
work together to make things happen as we will. You carry the pieces from place to
place, assembling them together in new ways and new contexts. But you always assume
that they will try to behave in new arrangements as they have tried to behave in others.
They will, in each case, act in accordance with their nature. (Cartwright 1999: 83)

Newton's physics and his conception of scientific methodology bear notable
similarities to those of Descartes. Not only are his three laws of motion loosely
modeled on Descartes', so too is his conception of the scientific method.[6] He
articulates his Four Rules of Scientific Reasoning (1686: Book III). There he
urges scientists to seek out those explanatory principles that are the most simple
(Rule 1), unifying (Rule II) and universal (Rule III).[7] Newton's physics is
predicated on the idea that the world operates always and everywhere through
the ministrations of a few simple unifying causal laws. Science strives to
articulate them.

We ended the introductory section with a puzzle: of Schrödinger's two
answers to the question 'What is Life?', why was only one ever really taken
up and assimilated into mainstream biology? Schrödinger's 'order-from-order'
answer suggests that the extraordinary capacities of organisms are vouchsafed
by the robust stability of their parts, in particular the chemical 'code script' out
of which they are assembled. This kind of thinking has been enormously
influential. It inspired the molecular revolution of twentieth-century biology.

---

[5] As we saw in Descartes' proclamation that in formulating his Principles of Physics, he considered 'only the various shapes and movements of [the universe's] parts'.

[6] There are profound differences of course. For one, Newton was famously scathing about
Descartes' conception of the relation between natural and mathematical phenomena.

[7] Rule IV stresses the importance of observation and measurement for scientific knowledge.

The alternative 'order-from-disorder' answer, by contrast, has been pretty peripheral to evolutionary thinking throughout most of its history. It is tempting to conjecture that at least a significant part of the reason for the differential success of Schrödinger's two answers is methodological. A commitment to the methodology of mechanism leads one naturally to seek an understanding of living things and processes in the properties and activities of their parts, governed by the laws of mechanics. Only the 'order-from-order' answer is amenable to any kind of productive and systematic investigation under the mechanistic methodology bequeathed to modern biology by the Scientific Revolution.

1.1   Contemporary mechanism

The methodology of mechanism that sparked the Scientific Revolution has received a potent update in recent years (Machamer, Darden and Craver 2000).

A mechanism is a structure performing a function in virtue of its component parts, component operations, and their organization. The orchestrated functioning of the mechanism is responsible for one or more phenomena. (Bechtel and Abrahamsen 2002: 26)

Contemporary mechanism doesn't just offer a methodology for science. It presents itself as at least a partial philosophical account of explanation: to explain is to cite mechanistic causes.

Causal processes, causal interactions, and causal laws provide the mechanisms by which the world works; to understand why certain things happen, we need to see how they are produced by these mechanisms. (Salmon 1984:132)

Just as in Descartes' and Newton's times, the explanation of the behaviour of complex entities, like the causation of their behaviour, is taken to be a bottom-up enterprise. Scientific explanation, on this view, proceeds in large measure by decomposition. Explaining the complex phenomena of the world, simply involves citing

the behavior of a system in terms of the functions performed by the parts and the interactions of these parts ... A mechanistic explanation identifies these parts and their organization, showing how the behavior of the machine is a consequence of the parts and their organization. (Bechtel and Richardson 1993:17)

Because of its reliance on decomposition as an explanatory strategy, mechanism lends itself readily to a kind of reductionism. One has to be careful here. In current philosophy of science the epithet 'reductionism' carries a stigma. Why this should be so is a bit of long story.[8] But 'reductionism' doesn't need to be saddled with these unsavoury connotations. All I mean by 'reductionism' in this

---

[8] Nicely told in different ways by Rosenberg (2006) and Brigandt and Love (2008).

context is that the methodology of mechanism gives us a good reason to suppose that an adequate explanation of the properties and activities of complex entities can be furnished by adverting to the properties and activities of their parts. This is a respectable and productive approach in biology as in most of the natural sciences.

At least in biology, most scientists see their works as explaining types of phenomena by discovering mechanisms, rather than explaining theories by deriving them or reducing them to other theories, and *this* is seen as reduction . . . . (Wimsatt 1974: 671)

One of the reasons that reductionism has incurred such odium in philosophy of science circles is that in the past it suggested, and indeed had been used to motivate, a kind of scientific imperialism, which undervalued the special sciences like psychology and biology, even threatened their existence. The imperialist thought went something like this. The sciences are arranged hierarchically in the sense that the respective domains of the special sciences are included in the domains of less special sciences as proper parts. Psychology, for example, ranges over cognitive beings, which (in these parts at least) are all organisms. So the domain of psychology is a proper part of the domain of organismal biology. But organisms are made of cells; so the domain of organismal biology is a proper part of the domain of cytology. But cells are aggregates of chemicals; so the domain of cell biology is contained within the domain of chemistry, and on down. The material world, after all, is a unity. At bottom, every fact is a physical fact, so every fact has a complete description (whether we know it or not) in the apparatus of physics.

Early reductionists, especially those in the grip of logical positivism, took this unity of the physical world to imply a unity of the sciences. If the respective domains of sciences are arranged hierarchically, and the phenomena at one level of organisation can be exhaustively accounted for by the phenomena at a more fundamental level of organisation, then, so the argument goes, we should expect that as science progresses, one by one the 'special' (less fundamental) sciences will be subsumed under, and superseded by, the less special. The endpoint of the process is the complete description of all natural phenomena in the vocabulary of physics. At that point, all other sciences will be parochial and otiose and will simply wither away. Oppenheim and Putnam (1958), for example, take it to be both an empirically observable trend within the sciences, and a guiding normative principle, that special sciences are one by one to be replaced by the lower sciences.

It is perhaps little wonder that this overzealous foundationalist reading of reductionism has found scant support. Not only does its normative claim suggest an invidious attitude to all scientific endeavours other than physics, its empirical claim has simply not been borne out (Fodor 1974). The special sciences, as they grow in their number, scope and success, show no inclination to be enveloped or subsumed by the more fundamental sciences.

Contemporary mechanists do not deny that there are special sciences. Typically, mechanists are perfectly sanguine about acknowledging stable and recurring regularities in the world that hold between complex entities (Craver 2007). Nor do they deny that explanations of these regularities can be given by citing the causal properties of the complex entities. Contemporary mechanism does not claim that the only *bona fide* mechanisms are the most fundamental (Bechtel 2006; Craver 2007). Nor does a mechanistic explanation cite only the components of system. A mechanistic explanation also relies on context, and a system's context typically isn't one of its components. Nevertheless, mechanism issues in a form of reduction. William Bechtel takes up the theme:

[M]echanistic explanations are commonly characterized as *reductionistic*. The notion of reduction that arises within mechanistic explanation, however, is very different from that which has figured either in popular discussions, or in recent philosophy of science ... in these discussions, appeals to lower levels are thought to deny the efficacy of higher levels ... Mechanistic reductionism neither denies the importance of context or of higher levels of organization ... The appeal to components in fact serves a very restricted purpose of explaining how, in a given context, the mechanism is able to generate a particular phenomenon. (Bechtel 2006: 40–41, emphasis in original)

These caveats notwithstanding, contemporary mechanism is 'reductionist' in a restricted and entirely reasonable sense: the causal capacities of a complex entity, in a context, can be exhaustively accounted for by the capacities and interactions of its parts.[9] In a given context, the capacities of a complex system are inherited from the capacities of its component parts. Many contemporary mechanists point out that in our explanatory practices we readily accept explanations that simultaneously advert to entities and activities at a variety of levels of complexity (Craver 2007; Craver and Darden 2013). Nevertheless, mechanism does appear to entail that for any explanation of a complex phenomenon, there is available (at least in principle) an exhaustive explanation that adverts to the component parts and processes of that phenomena. Furthermore, no explanatory loss is incurred by descending to the level of parts. There may, of course, be pragmatic reasons for not doing so. But if parts explain the capacities of wholes, there is always such an explanation to be had. This is a highly pluralistic, nonfoundationalist, stigma-free reductionism, but a reductionism all the same.

It is important to understand how profoundly productive this approach to understanding biological phenomena has been throughout the twentieth century. It has yielded an account of the most complex and recondite

---

[9] Mechanism in this sense is an example of what Winther (2011) calls 'part-whole science'. See also Griffiths and Stotz (2013).

organismal and evolutionary processes in terms of the mechanical interactions of complex molecules. In many ways, twentieth-century biology is the triumph of reductive mechanism.

1.2   Mechanistic reduction in evolutionary biology

Late nineteenth-century and early twentieth-century biology seems to have been something of a methodological/philosophical battleground. The problem of how to understand the phenomena of development and inheritance spawned a range of competing methodologies. As we saw in the Introduction, substance vitalists, like Driesch, took the organisation of development and inheritance to be under the control of some nonmaterial principle of organisation. Because of this, the dynamics of development and inheritance couldn't be revealed by attending only to the mechanical interactions amongst entities at *any* level of organisation. There were also, at the time, materialist holists, such as Wilhelm Roux, Ernst Haeckel, Claude Bernard and later Hans Spemann, who had no truck with the apparent anti-naturalism of the substance vitalists, but who nevertheless insisted that the mysteries of organismal development and inheritance could not be understood simply by attending to their component processes in isolation (Hamburger, Allen and Maienschein 1999).

What united all forms of holism was the clear recognition that living organisms were capable of activities that had no counterpart in the machine world: self-replication, purposeful or ordered response to stimuli, elaborate self-regulatory capabilities, and the incredible efficiency of their energy transduction. To understand these complex interactive processes, it was necessary to get beyond the individual parts to look somehow at the whole system or process. (Allen 2005: 267)

In contrast to these 'holists' there were also those, most notably the influential embryologist Jacques Loeb, for whom the objective of studying inheritance and development was to bring these processes under the ambit of physics and chemistry (Allen 2005). Loeb propounded a resolutely mechanistic, reductionistic conception of life.

The profusion of methodological approaches seems to have dwindled early in the twentieth century. Increasingly, mechanism came to predominate. Its record of success is sufficient to explain why.

1.2.1   Classical genetics

Thomas Hunt Morgan won the Nobel Prize in medicine in 1933 for his pioneering work on the chromosomal theory of inheritance. The research on chromosome structure generated by Morgan's laboratory, together with selective breeding experiments, combined to give us the classical conception of the gene. Morgan began his career as an embryologist under the guidance of the vitalist Hans Driesch. It was only after encountering Jacques Loeb, and coming to absorb the insights in Loeb's *Mechanistic Conception of Life* that Morgan turned from Driesch's holism to mechanism. That turn was decisive. As Garland Allen (2005) effectively documents, it was Morgan's embrace of mechanism, and his rejection of holism, that propelled him toward *both* his discovery of the classical gene, and the articulation of his chromosomal theory of inheritance.

For the holist embryologists, with whom Morgan began his career, the locus of the control of inheritance was not to be found within suborganismal parts, much less within cells. Rather, inheritance was the result of the orchestrated actions of the entire organism throughout its entire life. One implication of this view was that the processes of organismal development and inheritance were one and the same. As Morgan himself said (prior to his mechanist epiphany): 'we have come to look upon the problem of heredity as identical to the problem of development' (Morgan 1910: 449). He deemed the newly emerging 'particulate' theory of inheritance 'picturesque or artistic' (p. 460), but found it 'less stimulating for further research' (p. 460). For these reasons he was unsympathetic to the Mendelian conception of inheritance as controlled by discrete, suborganismal 'factors'. Upon encountering Johannsen's (1911) distinction between genotype and phenotype, and his conception of the gene, Morgan began to change his view.

When Wilhelm Johannsen coined the term 'gene' in 1909 he had little reason to suspect that genes might be discrete, isolable particulars. *Gene* was merely an instrumental concept. Johannsen famously 'framed no hypothesis' about its nature.

The word 'gene' is completely free of any hypothesis; it expresses only the certain fact, that many characters of organisms are somehow or other stipulated by the special . . . conditions . . . which are present in the gametes. (Quoted in Churchill 1974: 14)

Morgan and his coworkers discovered that specific differences in chromosomal structure correspond to differences in heritable characters. These specific regions became 'genes'. For Morgan and his colleagues, genes became mechanisms: 'entities' whose characteristic 'activities' formed the basis of inheritance and development. Members of the Morgan school were well aware, of course, that an enormously large and complicated plexus of processes was involved in the development of an organism's traits. Nevertheless they attributed a special role to genes in securing specific similarities and differences between organisms' inherited characters.

So it is that mechanism provided the methodological impetus for this conceptual shift from the instrumental gene, to the 'classical' gene. Morgan's embrace of mechanism led him to decompose organisms and to look at the workings of their parts independently. He sought self-subsisting, discrete entities whose characteristic activities could explain the phenomenon of inheritance. He found them in small, contiguous stretches of chromosomal material.

1.2.2   Molecular genetics

In 1953 Watson and Crick announced the discovery of the chemical structure of DNA (Watson and Crick 1953). Their objective was to demonstrate how the characteristic capacities of genes as units of inheritance are held in place by the capacities of the molecules of which they are composed. By demonstrating how nucleic acids are attached to a phosphate-sugar spine, by discovering how bases pair up, forming weak hydrogen bonds, they demonstrated how the mechanical relations amongst the parts of DNA enable it to 'carry out the essential operation of a genetic material, that of exact self-duplication' (Watson and Crick 1953: 96).

The discovery opened a path toward a new science of molecular genetics. Geneticists came to understand how, given the structure of DNA, alternative genes could cause the differences in phenotype associated with differences in classical genes. They also came to grasp the way in which a gene could act as both a template for its own replication *and* as a code for the production of a primary protein structure. That is to say, the structure of DNA furnished an understanding of how the 'molecular gene' could act as both the unit of inheritance and the unit of developmental control. This too is a wonderful example of a mechanistic reduction; classical genes get molecular explanations.

There is a uniform way of understanding the basic dogma of classical genetics at the molecular level. Differences in classical genes produce differences in phenotypes because they affect the action of molecular genes. (Waters 1994: 184)

It is important to observe that the molecularisation of the gene concept came about as a result of a general commitment of early twentieth-century biologists to mechanism. Molecular genetics arose out of a general push toward the understanding of biology as a fundamentally molecular phenomenon.

The molecularization of genetics did not arise directly from the experimental regime of classical genetics, ... It was rather part of an all-encompassing molecularization of biology that occurred in parallel with the development of classical genetics but had largely independent roots. (Müller-Wille and Rheinberger 2012: 162)

1.2.3   Gene regulation

Early developmental genetics was challenged by a version of the problem that vexed nineteenth-century developmental biology. If each cell contains the same genetic material, how is it that cells produce *different* tissues, structures and functions? If all the same genes were active in the same cells they should·all be pretty much the same. So different genes must be active in different cells. But, as

Morgan himself was quick to point out, if that should be so, it suggests that the ultimate controller of gene expression would be something *other* than genes (Keller 2000). This threatened the primacy of genes in development that had been one of great promises of molecular genetics. The central place of genes in development was salvaged by the discovery that what controlled and regulated the actions of genes was other genes. Jacob and Monod introduced a crucial distinction between structural genes and regulatory genes. In their discovery of the *lac* operon (1961) they were able to illustrate the way in which the mechanisms of gene transcription can be controlled by the mechanisms of gene regulation. The key to understanding differentiation, then, came to be the understanding of the way that genes turn other genes on or off, amplify and dampen gene activity.

These vignettes from three episodes of twentieth-century biology, scant though they are, nicely serve to emphasise the contribution made by the methodology of mechanism to the enormous growth of evolutionary biology. Contemporary mechanists are well aware that the recent history of biology attests to the efficacy of their chosen method.

[I]f one cannot appreciate this protracted and continuing discovery project, this project of discovering mechanisms, then one cannot fully appreciate why the working hypotheses of contemporary biology are worthy of our respect ... One cannot appreciate the Mendelian geneticist's understanding of heredity or the significance of the discovery of DNA for understanding how life works. One cannot understand biology or its history without understanding these protracted discovery episodes. And as most of these episodes are driven by the goal of discovering mechanisms, one cannot understand biology without understanding how phenomena are characterized and how mechanism schemas are constructed, evaluated, and revised. (Craver and Darden 2013: 10)

It is this same methodology that has inspired the molecularised order-from-order approach to life. At the same time it appears to have entrenched an extreme form of anti-exceptionalism: life is to be understood though the investigation of the principles of biochemistry (Morange 2009). The success of mechanism, however, says nothing about whether there might be a viable alternative.

1.3   Emergence

Emergence, or emergentism, is often taken to be the contrary of mechanism. If mechanism holds that the causal properties of a complex entity can be fully explained by the capacities of its parts, emergentism holds that the properties of complex entities play a *sui generis* explanatory role. We have encountered examples of various forms of emergentism in Aristotle's biology, and Kant's conception of organisms as natural purposes, in the varieties of vitalism. Sadly, these historical examples do little to advance the credentials of emergentism. Despite a certain pre-theoretic, intuitive appeal, emergentism has struggled for a compelling definition, and an accompanying methodology.

One way to interpret emergence is to see it as a metaphysical thesis: complex entities have causal properties that their parts do not. This may be interpreted as a wholly trivial claim or a quite substantive one. The trivial reading goes like this: Some complex entities have properties that could not be properties of the parts, but nevertheless are completely determined (and explained) by the parts. For example, gases have pressures; fluids have viscosities. However, this sort of observation falls well short of offering an alternative to mechanist reduction. Wimsatt captures the idea with characteristic precision and panache:

Philosophers commonly suppose that emergent properties are irreducible, but some rather nice things fall out of a reductive account of emergence ... Emergent phenomena like those discussed here are often subject to surprising and revealing reductionistic explanations. But such explanations do not deny their importance or make them any less emergent – quite the contrary: it explains why and how they are important. (Wimsatt 2000: 269)

If emergentism is to motivate (or necessitate) an alternative scientific methodology, it must say something more than that the properties of complex entities are different from, or surprising given, those of the parts.[10]

1.3.1 Ontological emergence

One very ambitious form of emergence makes the startling metaphysical claim that complex entities have causal powers that are *not* vouchsafed by the capacities and relations of the entity's parts. It seems obvious that were there to be such properties, then enumerating the properties of the parts, and their various combinations would be insufficient to capture all of the properties of the complex entities. Reductive mechanism would be blind to them. Gaps would be left in our understanding of the causal structure of the world that could only be filled by ascending to the level of complex entities and their properties.

Alas, according to an influential argument from Jaegwon Kim (1999, 2006) ontological emergence appears to be incoherent. Kim protests that ontological emergence requires a phenomenon he dubs 'reflexive downward causation', which he insists is nonsense.

Emergentism cannot live with downward causation and it cannot live without it. Downward causation is the raison d'etre of emergence, but it may well turn out to be what in the end undermines it. (Kim 2006: 548)

Kim's argument goes as follows. Suppose a complex entity, C, has the capacity to bring about effect E, but that capacity is not determined by the collection of

---

[10] That the properties of complex entities are surprising given the capacities of their parts is Chalmers' (2006) characterisation of 'weak' emergence.

C's parts, $<c_1, c_2, c_3, \ldots, c_n>$. In that event, Kim says, it would have to be the case that C also has the capacity to confer on $<c_1, c_2, c_3, \ldots, c_n>$ the power to bring about E. The reason is that C is nothing other than the collection of its parts, $<c_1, c_2, c_3, \ldots, c_n>$, so C could not bring about E unless $<c_1, c_2, c_3, \ldots, c_n>$ could, too. This is reflexive downward causation. It is incoherent, according to Kim, because when combined with a further reasonable metaphysical assumption, it entails a contradiction. That further metaphysical assumption goes by the name of the 'Causal Inheritance Principle' (CIP). It tells us that the properties of a complex entity are all a consequence of the properties of its parts and their arrangement. So C could not have a property unless the arrangement of its parts did as well. Conjoined with our initial assumption, then, the collection of parts must both have the causal power of the complex entity (by CIP) and *not* have that power (by the assumption). That is a contradiction.

If emergent properties exist, they are causally and hence explanatorily, inert and therefore largely useless for the purpose of causal/explanatory theories. If these considerations are correct, higher-level properties can serve as causes in downward causal relations only if they are reducible to lower-level properties. The paradox is that if they are so reducible, they are not really 'higher-level' any longer. (Kim 1999: 33)

Kim's argument against ontological emergence boils down to the assertion that it is incompatible with the Causal Inheritance Principle. Why should we not simply abjure the CIP in the face of the quite obvious evidence that complex entities have causal powers that the arrangement of their parts do not (O'Connor 1994)? Well, the Causal Inheritance Principle isn't so easy to renounce. It seems to be a straightforward consequence of the causal closure principle that we encountered in the introduction, conjoined with mechanism. Causal closure tells us that the causes of any physical phenomenon are physical. Mechanism tells us that the causes of a complex entity's properties are its parts. Ontological emergence seems doomed.[11]

1.3.2   Methodological emergence

Rather than suppose that complex entities have causal properties that their parts do not, some conceptions of emergence hold that the properties of complex entities enter into *explanations* of phenomena that cannot be explained by adverting to the parts alone. We might call this position 'methodological emergence'. Despite its lack metaphysical pretensions, it too has its problems. Here again, mechanism seems to militate against the supposition that the capacities of a complex entity could explain a phenomenon that could not be wholly and satisfactorily explained by the capacities of the parts.

---

[11] Something quite like it will emerge again in Chapter 10.

The reason is that mechanism appears to entail that explanation is transitive. The argument goes something as follows. It is plausible to suppose that the 'mechanism of' relation is transitive. If C is the mechanism that completely produces D, and D is a mechanism that completely produces E, then C is a mechanism that produces E. That being so, it should be easy to see how casting explanation as the citing of mechanisms should suggest that explanation is transitive. If $<c_1, c_2, c_3, \ldots, c_n>$ are the parts of some complex entity, C, and together they (and their activities) confer on C its causal powers, then if C is the mechanism that brings about some further effect E, then $<c_1, c_2, c_3, \ldots, c_n>$ must also bring about E. It seems, then, that everything that could be explained by adverting to C could also be explained by adverting to $<c_1, c_2, c_3, \ldots, c_n>$. The transitivity of explanation suggests that the properties of complex entities are explanatorily redundant.

Things might even be worse for explanatory emergence. There is a sense in which the explanation that adverts to $<c_1, c_2, c_3, \ldots, c_n>$ might supersede the explanation that cites C. The collection of parts $<c_1, c_2, c_3, \ldots, c_n>$ can explain why the complex entity, C, has the properties it has, yet, reductive mechanists believe, the converse relation does not hold. We cannot explain why the parts have their causal capacities by adverting to the whole.[12] So there seems to be nothing that can be explained by the activities of a complex mechanism, that could not be explained by the activities of its parts. Methodological emergence seems no better off than its metaphysically immodest alternative.

1.3.3   Emergent dynamics

Despite the difficulties that surround the articulation of a coherent emergent-ism, it has had some rather eloquent advocates. The 'British Emergentists' of the late nineteenth and early twentieth century were committed materialists who sought a corrective to the excesses of both materialist reduction and dualist vitalism (Garrett 2010). British emergentism resonated to the idea that the methodology that gave impetus to the scientific revolution is simply incapable of yielding an account of the supple dynamics of organisms.

Physical and chemical processes of a certain complexity have the quality of life. The new quality life emerges with this constellation of such processes, and therefore life is at once a physico-chemical complex and is not merely physical and chemical, for these terms do not sufficiently characterize the new complex which in the course and order of time has been generated out of them . . . The higher quality emerges from the lower level of existence and has its roots therein, but it emerges therefrom, and it does not belong to that level, but constitutes its possessor a new order of existent with its special laws of behaviour. (Alexander 1920: 46–47)

---

[12] I return to this claim in Chapter 10. It is not as secure as it may seem.

The sentiment is certainly laudable (and palpable), even if the details are wanting.

C.D. Broad (1925) attempts to develop a scientifically respectable materialist emergentism that makes good on the vague gestures. He begins with a critique of the prevailing conception of scientific methodology he calls Pure Mechanism. Pure Mechanism is marked out by four theses about the make-up of the world. There is . . . :

(a) a single kind of stuff,
(b) a single fundamental kind of change,
(c) a single elementary causal law, . . . and
(d) a single and simple principle of composition according to which the behaviour of any aggregate of particles, or the influence of any one aggregate on any other, follows in a uniform way from the mutual influences of the constituent particles taken by pairs. (Broad 1925: 45)

Broad thinks of Pure Mechanism as a kind of ideal. There are no pure 'Pure Mechanists'. Nevertheless these principles serve to identify some prominent working assumptions among scientists.

The most crucial thesis is the fourth. Something like it appears to underlie the mechanist strategy of causal decomposition; in order to understand a complex entity we take it apart and see what its components can do in isolation. It is this feature of the parts of a complex system to which Cartwright wishes to draw our attention: 'you always assume that they will try to behave in new arrangements as they have tried to behave in others' (Cartwright 1999: 83).[13] The assumption is that composition doesn't add anything to the capacities of the parts, or at least that the capacities of the parts are relatively unaffected by their contexts.[14]

Perhaps another way of making the same point is that in our mechanistic analyses, we apply a version of Mill's Method of Difference (Mill 1834). A simple version of the method of difference suggests that if you want to know what causal influence a part of a system has, you simply alter the part and see what difference your intervention makes to the system as a whole. But as Mill rightly notes, this procedure of causal decomposition is only appropriate where causes compose in an orderly and linearisable way, that is to say where the effect of multiple causes is their sum. Mill is careful to note that there is a catch: '[t]his principle. However, by no means prevails in all

---

[13] It is a surprisingly widely (if tacitly) made assumption in evolutionary biology. It appears to motivate Elliott Sober's (1984) claim that the influences on population change, like selection and drift, can be decomposed in the manner of Newtonian forces, and in Okasha's (2006) supposition that statistical decompositions of within-group and between-group change can be resolved into causes operating at distinct levels of organisation.

[14] This certainly seems to be an implicit assumption in Kim's argument (above), for a discussion of which see Ganeri (2011).

departments of the field of nature' (1843: 243).[15] In particular, it breaks down in biology.

> All organized bodies are composed of parts similar to those composing inorganic natures ... ; but the phenomena of life which result from the juxtaposition of these parts in a certain manner bear no analogy to any effects which would be produced by the action of the component substances considered as mere physical agents. To whatever degree we might imagine our knowledge of the properties of several ingredients of a living body to be extended and perfected, it is certain that no mere summing up of the separate actions of those elements will ever amount to the action of the living body itself. (Mill 1843: 243)

The method of difference works only if the causal contributions of all the parts of a system are sufficiently context-insensitive. That is to say that in changing one part of the system does not elicit any compensatory changes in the other parts.[16]

Many systems are organised in such a way that the various parts have highly context-sensitive 'nonlinear' and 'cyclical' causal relations. Contrary to the presumptions of the analytic method, they do not 'behave in new arrangements as they have tried to behave in others'. The activities of one component of the system alters the activities of other components, which frequently, in turn, redound upon the first component, through a complex feedback loop. In systems like this there is a sort of causal reciprocity between part and whole. The way that a part affects the whole can only be understood in the context of the way that the entire system affects, controls and regulates the activities of the parts. The strategy of causal decomposition fails because the causal contribution of a component to the dynamics of the system cannot be understood in isolation from the system as a whole. Indeed, this sort of cyclical, nonlinear, causal structure is the norm in biological systems (Camazine et al. 2001).

Vivid examples are to be found in gene regulatory networks. Genes typically do not work as isolated agents, but as parts of complex self-regulating systems. The discovery of gene regulatory networks arose in part out of the failure of traditional mechanist approaches to the study of gene action (Gibson 2002). The usual way to understand what a gene does is to remove or disable it and see what happens. It turns out that 'knocking out' a gene in a gene-regulatory network often has no overall effect. But this doesn't mean that the target gene makes no causal contribution to the activities of the system in normal circumstances. The phenomenon is a reflection of the fact that the relations among the components of a gene regulatory system are nonlinear, cyclical and

---

[15] Typically, emergentists (e.g., Mill, Broad, Engels) thought that any thorough explanation of the properties of complex molecules would also outstrip the methods of mechanism.

[16] I take up some of the implications of the assumption of composition of causes in Chapters 5 and 10.

self-regulating (Gibson 2002; Von Dassow and Munro 1999). An interference to one part of the network ramifies, causing changes to all the other parts. These changes are compensatory; they permit the system as a whole to maintain its function, despite the perturbation.

The capacity of regulatory networks to compensate in this way is called 'distributional robustness' (Wagner 2007), and it is no mere exotic quirk. It is integral to the stability, growth and function of organisms.

Living things are unimaginably complex, yet also highly robust to genetic change on all levels of organization. Proteins can tolerate thousands of amino acid changes, metabolic networks can continue to sustain life even after removal of important chemical reactions, gene regulation networks continue to function after alteration of key gene interactions, and radical genetic change in embryonic development can lead to an essentially unchanged adult organism. (Wagner 2007: 176)

Distributional robustness consists in the capacity of a system to regulate the activities of its parts.

When one part fails or is changed through mutations, the system can compensate for this failure, but not because a 'back-up' redundant part takes over the failed part's role. (Wagner 2007: 177)

Distributional robustness seems to call for an alternative methodology, in which the dynamics of the entire system are integral to an explanation of the activities of its parts. The component genes in a gene regulatory system, for example, behave in the way that they do precisely because their behaviour is regulated by the system as a whole. Each gene in each regulatory system has a repertoire, a range of activities in which it could partake. Which element of its repertoire is activated in a given context depends on its context. But the relevant context here is the robust, stable activity of the whole system.

This sort of adaptive dynamics seems to be the rule in biology; gene regulatory networks, genomes, cells, tissues and entire organisms exhibit this kind of complex, adaptive, nonlinear dynamics. Scientists and philosophers are beginning to realise that new conceptual tools are required.[17] Increasingly they are looking toward emergence as an alternative to reductive mechanism.[18]

Taking account of nonlinear dynamics and feedback causal processes in addition to static and linear representations, which may be adequate for simpler domains of nature, is one of the ingredients of understanding complexity. Emergence identifies an important class of phenomena, whose analysis permits a reframing of some of the standard approaches to understanding scientific explanation. (Mitchell 2012: 184)

[17] This movement is sometimes referred to as the re-emergence of emergence (see Clayton and Davies 2006).
[18] See contributions in Henning and Scarfe (2013).

There is a two-way influence between these systems and their parts that evades traditional mechanism. We can explain as much about the regular activities of the parts by citing the way they are constrained and regulated by the whole system, as we can explain the activities of system by citing the activities of their parts.

René Thom recommends that biologists embrace a version of 'Structuralism' that he adapts from the anthropological theories of Claude Lèvi-Strauss to explain the dynamics of complex entities like organisms and fluids: 'the fine structure, molecules for fluids, cells for animals, is practically irrelevant for understanding the global regulatory figure . . . of the total system' (1972a: 88). He resists the very idea that emergence or methodological vitalism requires any special metaphysical pleading. On the contrary, the mechanist,

postulates a reduction of facts about living processes to pure physiochemistry, but such a reduction has never been experimentally established . . . We must reject this primitive and almost cannibalistic delusion about knowledge, that an understanding of something requires that we first dismantle it . . . . (Thom 1972b: 158–159)

In contrast to the predominant mechanistic atomism of much of the Modern Synthesis, biologists have increasingly come to regard evolution from the viewpoint of organisms as entire systems. The focus of attention is on the dynamics of entire complex, self-regulating systems that constantly exchange matter and energy with their surroundings. These are systems that have the capacity to implement features of their own repertoires in response to the conditions they encounter.

[W]e have witnessed a paradigm shift in scientific thinking from an atomistic, mechanical, reductionist viewpoint to a systems perspective that incorporates cell circuitry and molecular networks into a more integrated view of cell and organismal activities . . . . (Shapiro 2011: 129)

This shift consists in a change of explanatory perspective. The new perspective accords explanatory primacy to the dynamics of whole systems over the capacities of their parts. This is a form of emergentism (Shapiro 2011), to which we shall return in Chapter 10.

Emergentism is now a renascent approach to the understanding of complex systems. It is still in its infancy, and still striving for a definitive articulation. Nevertheless, it at least offers the prospect of an alternative to the methodology of mechanism that has drawn researchers' attentions away from organisms as dynamic, self-organising, adaptive entities, and focused attention on organisms as mere aggregates of mechanical parts.[19]

---

[19] As Stephen Jay Gould quipped: 'Organisms are stuffed full of emergent properties' (2002: 620).

Conclusion

Twentieth-century biology witnessed unprecedented growth. From the discovery of the structure of DNA, to the sequencing of genomes, the understanding of protein synthesis, the biochemistry of metabolism, twentieth-century biology has encountered singular success in revealing the molecular mechanisms that underpin biological processes, including evolution. Biological understanding has flourished under the guidance of the 'molecular revolution'. Along the way, it gave us a gene-centred conception of evolution.

It is important to appreciate how much this enormous growth in biological knowledge is beholden to its chosen methodology. The intensive pursuit of mechanism has yielded a fantastically successful, yet comprehensively anti-organismal gene-centred biology. But just as the molecular concept of the unit gene inspired a research programme that was ultimately to undermine its own precepts, mechanism may have revealed a set of phenomena whose proper treatment outstrips its own austere explanatory resources.

2 Ensemble thinking
Of struggle and abstraction

Legend has it that, upon reading a prepublication manuscript of the *Origin of Species*, the great Victorian biologist T.H. Huxley proclaimed: 'How exceedingly stupid not to have thought of that'. Huxley's reaction is at once quaint and faintly shocking, but it also reveals much about Darwin's theory. Huxley was not being churlish or disparaging, one supposes; nor was he attempting to diminish Darwin's intellectual achievement. On the contrary, he was implicitly acknowledging the peculiar kind of acuity required to formulate such a position.

The theory of natural selection, as it appears in the *Origin*, is startling for its metaphysical modesty. Prior to Darwin, most theories of the fit and diversity of organic form posited some extra causal power or explanatory factor – be it an entelechy, a *vis essentialis*, a substantival form or a deity – that could not be directly observed in the day-to-day activities of organisms. Darwin's conceptual breakthrough came in his realisation that no such extra factor is needed. The daily activities of organisms really are sufficient by themselves to explain fit and diversity, after all. Darwin saw that the question of fit and diversity should be posed as a question about the constitution of populations: how do populations come to comprise individuals so well suited to their conditions of existence.[1] Once the question is framed in that way, a keen natural historian's eye can readily see that 'all these results *follow inevitably* . . . from the struggle for life' (Darwin 1859 [1968]: 114). It is this discovery – that population changes 'follow inevitably' – that makes Huxley's remark so apposite.

It is perhaps instructive in this regard to compare Huxley's response to Darwin's theory with the general reception of Einstein's general relativity. No one, to my knowledge, said 'how exceedingly stupid not to have thought of *that*'. On the contrary, when the British physicist Sir Arthur Eddington was asked whether he was one of only three persons in the world who understood the theory, he quipped 'I'm trying to think who the third one is'. Einstein's theory of general relativity, for all its incandescent genius, certainly lacks the

---

[1] In changing the focus of the question of fit and diversity from individual organisms to populations, Darwin is introducing population thinking into biology (Mayr 1963).

metaphysical and theoretical simplicity of Darwin's. Indeed it makes some extravagant commitments. It posits spacetime as a substance, variable in its structure, described by non-Euclidean geometries, that simultaneity is frame relative, that the speed of light is limiting. Formulating General Relativity is not a matter of finding out that after all the theoretical resources we need are already to hand.

Darwin's thinking establishes that the study of fit and diversity is, in some sense, the study of the dynamics of populations. The population, rather than the individual, becomes the object of interest in evolutionary theorising. As evolutionary theory expanded with the advent of the Modern Synthesis, the theoretical tools for studying population dynamics became increasingly sophisticated. At each stage – the formulation of Darwin's theory, and the inception of the Modern Synthesis – individual organisms relinquish some of their explanatory importance to evolution. In Darwin's evolutionary thinking, organisms surrender some, but by no means all, of their theoretical significance. It is not until we ascend to the Modern Synthesis that we see the complete alienation of organisms from evolutionary theory. If the theoretical core of evolutionary biology, theoretical population dynamics, doesn't need to take into account the properties of organisms, then it is hardly surprising that organisms should have slipped from our view as evolutionary theory grew and developed throughout the twentieth century.[2]

2.1   Natural selection: Cause or effect?

It is customary to think that in discovering the process of natural selection, Darwin has also identified the cause of descent with modification. To be sure, there is some exegetical support for this interpretation. Darwin introduces natural selection by way of a causal analogy. Chapter 1 of *The Origin* discusses in detail the ways in which captive populations can change in remarkable ways under the guidance of selective breeding. A breeder selects which individuals mate with whom, which survive and which perish. These interventions, Darwin notes, eventuate in quite extensive changes in the lineages being bred. Selective breeding produces such diverse varieties that were we to discover them in nature we would almost certainly take them to be different species.

Darwin then asks, rhetorically, if artificial selection could produce this range of varieties in just a few generations, what could natural selection do working incessantly, on all aspects and phases of organisms' lives over thousands and millions of years. Indeed he compares the respective efficacies of

---

[2] My thinking about these issues has been greatly influenced by various works by, and discussion with, André Ariew.

artificial and natural selection, deeming the latter to be more comprehensive, nuanced and powerful.

As man can produce and certainly has produced a great result by his methodical and unconscious means of selection, what may not nature effect? Man can act only on external and visible characters: nature cares nothing for appearances, except in so far as they may be useful to any being. She can act on every internal organ, on every shade of constitutional difference, on the whole machinery of life. Man selects only for his own good; Nature only for that of the being which she tends. Every selected character is fully exercised by her; and the being is placed under well-suited conditions of life ...; Can we wonder, then, that nature's productions should be far 'truer' in character than man's productions; that they should be infinitely better adapted to the most complex conditions of life, and should plainly bear the stamp of far higher workmanship? (1859 [1986]: 83–84)

Passages like this suggest that Darwin takes the processes of natural and artificial selection to differ not in kind, but only in degree. Artificial selection is most definitely a cause of population change. It is a process that is imposed upon populations from outwith and brings about changes in population structure. So, by analogy, one might suppose that natural selection is the same sort of cause of population change too. Certainly Darwin often speaks of natural selection in causal terms. The passage quoted above continues:

It may be said that natural selection is daily and hourly scrutinizing, throughout the world, every variation, even the slightest; rejecting that which is bad preserving and adding up all that is good; silently and insensibly working, whenever and wherever opportunity offers, ... (p. 84)

In normal circumstances, to say that one thing is a cause of another, to say that P causes Q, carries certain counterfactual implications. Typically, it suggests that *holding everything else fixed*, if P hadn't happened, then Q wouldn't have either. A causal reading of the foregoing passage, then, would seem to imply that were selection not to 'scrutinise' a population in the way it does, were it not to reject the bad and preserve the good, then (holding everything else fixed) the population would not change in the same ways. That reading, in turn, invites the supposition that selection is a process over and above the aggregate of individual organisms' struggles to survive in their conditions of existence. Specifically, it represents natural selection as something extraneous to, superadded to, the struggle for existence, in just the way that artificial selection is.

If this interpretation is right, then Darwin's brilliant insight is mistaken; the struggle for existence in varying populations is *not* sufficient to cause fit and diversity. On this view Darwin really has introduced a new theoretical posit, or discovered a new process, wholly other than the aggregate of individual organisms' lives, deaths and reproductions. Tempting as the causal interpretation of selection is, it does little justice to Darwin's proclamation that 'all these things, ... *follow inevitably* from the struggle for existence' (emphasis added).[3]

The 'following inevitably' claim – not to mention Huxley's 'how exceedingly stupid' remark – suggests a different interpretation. Natural selection is in a sense spontaneous. Given the normal activities of organisms, nothing needs to be added to get populations to change in the ways that Darwin describes as natural selection. On this view, Darwin's claim that 'all these things, ... *follow* inevitably' signifies that natural selection is an analytic consequence of the lives, deaths and reproduction of individuals.[4]

2.1.1   Analytic consequence

This idea of an 'analytic consequence' needs a little clarification. Consider an example.[5] A particular system comprises two particles, p_1 and p_2, in a container. The particles are moving away from each other at equal and constant velocity, in such a way that their centre of mass, c, remains at a constant location. Now imagine that one of the particles, p_1, strikes the wall of the container, while p_2 continues on its way; p_1 bounces back the way it came. Immediately on p_1's change in velocity, the centre of mass, c, shifts (it moves in the direction of p_2 and away from p_1).

The movement of c is a special kind of event. It is a consequence of the change in velocity of p_1 (or better, the change in the change in the *difference in velocity* between p_1 and p_2), but it is instantaneous. Even though c is set in motion by the change in velocity of p_1 there is no signal sent from p_1 to c, to initiate its movement. There couldn't be, it would have to be propagated faster than the speed of light, and we are told that no signal could be. But no such signal is needed. As the relative locations of p_1 and p_2 change, so does the location of their centre of mass, spontaneously. The locations (and the masses) of p_1 and p_2 *entail* the location of c. This is the sense in which one event – the movement of c – can be an analytic consequence of the occurrence of another.

2.1.2   Selection as a higher order effect

On the analytic consequence interpretation, natural selection is not a cause of population change; it is a 'higher order effect'. A higher order effect is the effect on an ensemble of causal processes acting at the level of its

---

[3] One may attempt to preserve this reading by positing an environment, or as Lyell had it, an Economy of Nature, as a unified entity wholly external to organisms that is the principle cause of evolutionary changes in form. I take this line of reasoning up in Chapter 10.

[4] Neil Tennant (2014) elegantly, and rigorously, captures the 'analytic' nature of Darwin's theory.

[5] This example comes from Matthen and Ariew (2009).

individual components. Here again, an example may help. If we place a drop of concentrated potassium permanganate into a beaker of water, we observe that the drop diffuses. Diffusion is a result of the movement of individual permanganate molecules in the solution. The several movements of these molecules are random, but their collective net effect is highly ordered and robustly predictable. How do we explain how aggregate order arises from individual random motion? Significantly, we do not need to invoke an ensemble level causal process in order to explain the changes in the solution as a whole.

Erwin Schrödinger (1944) offers the following account. Suppose we were to place a membrane into the permanganate solution some time before it has reached equilibrium, and to observe the collisions between the membrane and the permanganate molecules, we would find that more particles collide with the membrane from the high concentration side than from the low. That is perfectly natural and obvious, as there are many more molecules on the high concentration side. So, even assuming that the molecules are moving randomly, we should expect a higher frequency of molecules traveling from high concentration to low than in the reverse direction. What this means in terms of the solution as a whole is that, given these molecular motions, gradually it will spontaneously move toward an equilibrium in which the permanganate molecules are equally distributed throughout the solution. The ensemble-level process is diffusion. Diffusion is simply an effect – a higher order effect – of the movements of individual molecules. The change in concentration is, like the change in the location of the centre of gravity, an analytic consequence of the movements of molecules.

In the same way, the process that Darwin discovered is a higher order effect. It is simply the effect on a population of the effect on individuals of all those myriad causes of living, dying and reproducing. Population changes of a regular and predictable sort – increasing diversity of populations, and increasing fit of organisms – follow inevitably, even though the details of individual lives may be stochastic and highly unpredictable. In an important sense, natural selection is causally redundant (Brunnander 2007). It is, never-theless, like passive diffusion, a real and, most assuredly, a significant phe-nomenon. Darwin's masterstroke, then, is not like that of Einstein's in which an entirely new theoretical apparatus, replete with radical metaphysical posits, theoretical quantities, bold idealising assumptions, is brought to bear in the solution of a persisting and intractable set of problems. Rather it resides in the perspicacity of recognising that all the resources required to solve the 'mystery of mysteries' are already at hand. Nothing new needs to be added. All that is required is a simple change of perspective. How exceedingly stupid not to have thought of that!

2.2   Two grades of population thinking

Ernst Mayr (1975) dubbed the required new perspective 'population thinking'.[6] Sober (1980), following Mayr, offers an influential characterisation.

Darwin ... focused on the population as a unit of organization. The population is an entity, subject to its own forces, and obeying its own laws. The details concerning the individuals who are parts of this whole are pretty much irrelevant. Describing a single individual is as theoretically peripheral to a populationist as describing the motion of a single molecule is to the kinetic theory of gases. In this important sense, population thinking involves ignoring individuals. (Sober 1980, emphasis in original. Quoted from Sober 2006: 344)

Sober is certainly correct to say that the diagnostic mark of population thinking involves taking the population to be an entity whose properties are to be explained. It is true too, in a certain restricted sense, that details of individual activities are 'pretty much irrelevant' to Darwin's account of population change. But Sober's characterisation is ambiguous. It risks conflating two distinct grades of population thinking. The distinction is important because it corresponds to the difference between Darwin's theory and its Modern Synthesis successor. Failing to attend to the distinction can obscure the quite pronounced divergences between these two versions of evolutionary theory. Having a clear grasp on the difference will help us to understand the magnitude of the transition from Darwinian natural selection to Modern Synthesis natural selection. That, in turn, will help will illustrate the way in which the Modern Synthesis further marginalised organisms. I shall call these two grades of population thinking 'Malthusian' and 'Galtonian' after two of their most prominent exponents.

2.2.1   Malthusian population thinking

Darwin cites the influence of the English political scientist Thomas Malthus as seminal in his discovery of natural selection. Malthus' prime concern lay with social conditions among the English working class. His influential publication of *An Essay on the Principle of Populations* (Malthus 1798) identified a structural challenge that any population of human beings must face. Human populations, Malthus argued, tend if unchecked to grow at a geometric rate of increase. Given that, any population will outstrip the resources available to it in very short order. The result, according to Malthus, will be dearth, destitution and deprivation all round. In order to avoid this dire consequence, Malthus argued that some check needs to be applied to humans' natural proclivities to reproduce.

---

[6] This section draws heavily from André Ariew's (2008) and from numerous conversations with him. I am exceedingly grateful to André for his help with these issues.

Certainly Darwin was greatly impressed by the Malthus' argument from fecundity. He provides some engaging hypothetical examples of his own.

> The elephant is reckoned to be the slowest breeder of all known animals, and I have taken some pains to estimate its probable minimum rate of natural increase: it will be under the mark to assume that it breeds when thirty years old, and goes on breeding till ninety years old, bringing forth three pair of young in this interval; if this be so, at the end of the fifth century there would be alive fifteen million elephants, descended from the first pair. (Darwin 1859 [1968]: 84)

Where Malthus sees misery as the consequence, Darwin sees harmony, the 'fit' of organisms to the conditions of existence. Given that natural populations vary, Darwin surmises, the consequence will not be despair, but differential success in the struggle, and increased adaptedness. Despite their diametrically opposed conclusions, Darwin is applying the same strategy as Malthus. They are both drawing inferences concerning the consequences to a population of individuals acting in the way they do.

The causes of each individual organism's death, the conditions of its birth, the circumstances of its survival and reproduction may be unpredictable, and highly variable, subject to all manner of vagaries, like the motions of those permanganate molecules in solution. Nevertheless, despite all this low-level complexity, this motley of multifarious causes resolves itself into highly regular, predictable trends discernible only at the level of populations. 'So, from Malthus, Darwin learns to understand how multiple causes might conspire to create large-scale regularities in fixed law-like ways (population phenomena)' (Ariew 2008: 81).

To be sure this is population thinking. It involves ignoring individuals insofar as one does not have to enumerate the properties of each individual severally in order to explain the population-level effect. Nevertheless, there is a sense in which individuals are indispensable to this way of thinking. The population effect is explained as the aggregate of individuals' activities, in just the way that diffusion is explained as the aggregate of individual molecular motions.

2.2.2   Galtonian population thinking

A more thoroughgoing brand of population thinking can be found in the work of Darwin's second cousin Francis Galton. Galton had a particular interest in inheritance – which he took to be a pattern of resemblance between ancestors and descendants – in human populations. He developed a battery of statistical tools in part as a set of methods for studying this phenomenon.

The pattern of inheritance of quantitative, continuously varying traits, like intelligence or height, was a paradox to Galton. It defied common sense. Galton noticed that offspring tend to have values for these sorts of traits that are intermediate between those of their parents. Extremely tall individuals, for example, are rare. It is overwhelmingly likely, then, that any mate of such a tall person would be less exceptionally tall. The offspring of such a mating would tend to have an intermediate 'mid-parent' value, and so individuals of this exceptional height should be expected to decrease in frequency generation on generation, a phenomenon he called 'regression to the mean'. Given that, Galton reasoned, we should expect that rare, extremal traits should gradually disappear from a population. And yet, to Galton's bewilderment, they don't. The distribution of these traits remains relatively constant generation upon generation.

Galton initially toyed with the idea that there must be an extra force acting on the population to counteract the loss its extremal traits. It is understandable why. The aggregate activities of individuals do not appear sufficient to account for the population regularity. Something else is needed. He even entertained the notion that this force might be natural selection. Eventually, however, he realised that no such extra factor is required, that the reproduction of individuals should after all be expected to maintain a constant distribution of traits. The clue to this resolution lay in the statistical properties of populations.

Galton reasoned as follows. The traits of offspring will be normally distributed about the mid-parent value. Parents with extremal values for a particular trait will tend to produce offspring whose average value for the trait in question will be less extreme. Crucially though, a very small proportion of offspring of parents with less extreme values will also produce some small number of offspring with extremal values. For example, while the offspring of an exceedingly tall parent will be on average shorter, shorter parents will tend to have a very small proportion of very tall offspring. Now, the shorter parents are much more numerous than the exceptionally tall parents. So, even though the proportion of exceptionally tall offspring that shorter parents produce is minute, it is enough to replenish the number of exceedingly tall individuals in the population lost through regression to the mean. An equilibrium is struck between regression and replenishment. In this way generation upon generation the distribution of these traits remains constant.[7]

The important point for our purposes is that Galton's reasoning exemplifies a distinctive grade of population thinking. Galton seeks to explain a regular phenomenon of populations – the stable distribution of inherited traits – by adverting to *a statistical property of populations*.[8] Galton's version of population thinking, then, contrasts with Malthus' in that the latter but not the former needs recourse to the properties and activities of the individual members of the

[7] See Ariew, Rice and Rowher (2015) for an illuminating discussion of Galton's reasoning.
[8] This is an example of what (Lange 2013) calls 'really statistical explanations'. See also Walsh (2014a) and Ariew, Rice and Rohwer (2015).

populations. All Galton needs is the *distribution* of those properties. In a very strong sense, Galton's population thinking really is '... about ignoring individuals'. As in Malthusian population thinking, individuals are no part of the explanandum. But unlike the Malthusian version, in Galton's population thinking, individuals are no part of the explanans either.

With the distinction between Malthusian and Galtonian grades of population thinking in hand, we can identify the ways in which Darwin's theory of natural selection does and doesn't exemplify population thinking. Darwin is a Grade 1, Malthusian, population thinker. The phenomenon to be explained, Darwin correctly notices, is a property of populations. But because selection is identified as a higher order effect, it is to be explained by citing the properties and activities of individual organisms, the struggle for life.

2.3   Grade 1 population thinking

The metaphysical modesty of Darwin's theory that provoked Huxley's exclamation may be its biggest selling point. It demonstrates that if we accept that populations vary, that offspring resemble their parents, and that organisms struggle for life, then we are also committed to natural selection. The same metaphysical modesty may also be its principal weakness. Darwin's theory is severely limited in the degree to which it can explain, predict and measure evolutionary change. Nowadays, evolutionary change is thought of as a change in the trait structure of a population. Trait types become comparatively more or less frequent under natural selection and drift. That being so, the apparatus of Darwin's theory doesn't actually give us a criterion for saying when evolution has occurred. Nor does it provide a way of quantifying or predicting the degree of evolutionary change in a population. These are surely things that we would like our theory to enable us to do.

The deficiencies of Darwin's theory are easily demonstrated. These inadequacies also motivate the ascent to Grade 2 population thinking, found in the successor to Darwin's account of natural selection – the population dynamics that forms the core of the Modern Synthesis.[9]

Suppose we have a population of asexually reproducing individuals. There are three trait types of interest, with two variants each: Size (large or small), Shape (square, triangle), Colour (dark, light). In all, then there are eight combinations of traits, total phenotypes that an individual might have (see Table 2.1 below). We can assign a fitness (an expected number of offspring) to each individual. The average fitness for each of the eight total phenotypes is given in Table 2.1.

---

[9] This discussion is taken primarily from Walsh (2004).

Model 1

Table 2.1 *Individual fitnesses frequencies (Model 1)*

Individual		Individual Fitness	Frequency F_0	Frequency F_1
1		1.2	100	120
2		0.8	100	80
3		1.2	100	120
4		0.8	100	80
5		0.8	100	80
6		1.2	100	120
7		0.8	100	80
8		1.2	100	120
		Total	800	800

Table 2.1: Heritable variation in individual fitnesses for a population. Individual fitness is the expected reproductive output of each individual.

Now we ask whether this population is undergoing evolution by natural selection. There is certainly heritable variation in the ability of organisms to leave offspring in the struggle for existence, as evidenced by the variation in fitness. The population certainly changes in its constitution, in that as some lineages become more prominent others become less so. The conditions that Darwin stipulates typically cause natural selection obviously obtain here.

But if our interest is whether this change of population structure is a genuine *evolutionary* change, that is to say change in trait structure, the table will not tell us. Does *Big* increase in frequency with respect to *Little*, for example? The table doesn't say. Of course, this is easily figured out. We just tally up the starting and finishing frequencies for the each of the trait types. These are given for this population in Table 2.2.

Table 2.2 shows that this population is not undergoing evolutionary change. There is no change in the relative frequencies of any of the trait types. And it is easy to see why; there is no variation in trait fitnesses. That is to say there is no variation in the growth rates of the abstract trait types.

Now consider a similar model population, with the same degree of variation in individual fitness, as in Model 1, given below.

Table 2.2 *Trait fitnesses and trait type frequencies (Model 1)*

	Trait Types					
	Big	Little	□	△	⌇	⌇
Trait Fitness	1.0	1.0	1.0	1.0	1.0	1.0
Frequency F_0	400	400	400	400	400	400
Frequency F_1	400	400	400	400	400	400

Table 2.2: Starting and finishing frequencies, and fitnesses of the eight trait types. Trait fitnesses are here calculated as the average reproductive output for individuals with a particular trait.

Model 2

Table 2.3 *Individual fitnesses and frequencies (Model 2)*

Individual		Individual Fitness	Frequency F_0	Frequency F_1
1	▲	1.2	100	120
2	▲	0.8	100	80
3	■	1.2	100	120
4	■	0.8	100	80
5	◻	0.8	100	80
6	◻	0.8	100	80
7	△	1.2	100	120
8	△	1.2	100	120
		Total	800	800

Despite the fact that the degree of individual variation is the same here as in Model 1, the evolutionary outcome is different. Of course that cannot be discerned simply by inspecting Table 2.3. We must tabulate the starting and finishing frequencies of the trait types in this population, just as we did in Table 2.2, Model 1.

Here we see genuine evolutionary change; there is a change in the relative frequencies of the trait types. *Big* increases with respect to *Little* and *Triangle* increases relative to *Square*. These changes, moreover, are predicted and explained by the trait fitnesses.

Table 2.4 *Trait fitnesses and trait type frequencies (Model 2)*

	Trait Types					
	Big	Little	$\square$	$\triangle$	$\mathbf{z}$	$\mathbf{z}$
Trait Fitness	1.1	0.9	1.1	0.9	1.0	1.0
Frequency F_0	400	400	400	400	400	400
Frequency F_1	440	360	440	360	400	400

Table 2.4: Starting and finishing frequencies, and fitnesses of the eight trait types. Trait fitnesses are here calculated as the average reproductive output for individuals with a particular trait (as in Model 1).

The transition from *individual* fitnesses in the first table for each model to *trait* fitnesses in the second table is arithmetically trivial. But it is conceptually significant. Individual fitness is a causal property of a concrete entity, an organism. Trait fitness is a property of an abstract entity, a trait type. It is represented in this population by a group of individuals, those possessing a token of the trait type in question. Trait fitness in turn, is a statistical property, here represented as the average reproductive output of a group of individuals (again, those with the trait in question).[10]

These models, simple as they are, demonstrate two things. The first is that the sort of Malthusian population thinking employed by Darwin, does not render a criterion of evolutionary change. Nor does it allow us to predict or quantify the degree of evolutionary change. Its deficiency lies in the very metaphysical modesty that makes it so intuitively compelling. Evolution may be the consequence of individuals struggling to survive, but citing these individual-level causes of population change is insufficient to explain the object of natural selection theory: change in trait structure. In this respect, Darwin's conception of natural selection is inadequate as a theory of the dynamics of populations. The second moral of these simple models is that in order to explain the dynamics of populations, we must invoke something else: the distribution of trait fitnesses. Invoking trait fitnesses, in turn, requires us to ascend to an ontology of abstract entities, trait types and the statistical properties of a population.

An adequate theory of evolutionary population dynamics, then, requires us to abandon Darwin's Malthusian population thinking, and adopt an altogether more comprehensive, Galtonian, version. This, in fact, is precisely what

---

[10] More sophisticated measures of fitness are more obviously statistical. They involve the mean and variance of reproductive output, and population size. (See Gillespie 1977; Walsh 2010a, 2014a.)

occurred in the transition from the Darwinian conception of natural selection to the Modern Synthesis conception.

2.4   Natural selection in the Modern Synthesis

By the early twentieth century, evolutionary biology had become factionalised. There were two competing camps, neo-Darwinians and Mendelians, and the exchanges between them were famously unseemly and rancorous.[11] Neo-Darwinism was a theory of adaptation. It held that adaptive evolution occurred by the gradual accumulation of very small improvements through the process of Darwinian natural selection. Mendelism, in contrast, was primarily a theory of lineage change and stasis. It proposed that evolutionary changes tended to occur as large scale, discontinuous saltations. Mendelians cited the fact that inheritance is mediated by the transmission of 'factors' that in most cases were passed unaltered from parent to offspring. Differences in Mendelian factors between individuals were thought to be manifest as discontinuous and discrete differences in form. Evolutionary change occurs through the introduction of new factors via the process of mutation. The Mendelian theory of inheritance suggested that inherited differences between individuals are too coarse-grained and discontinuous to provide the minute and graded heritable differences between organisms required by the neo-Darwinians. These two approaches to evolution seemed to be irreconcilable.

R.A. Fisher set out to demonstrate that Mendelian inheritance is compatible with Darwinian selection (Fisher 1918). The outcome of Fisher's endeavour to reconcile neo-Darwinism with Mendelian inheritance was the Genetical Theory of Natural Selection, the theoretical engine of the Modern Synthesis theory of evolution. Fisher was inspired by the statistical treatment of thermo-dynamics, and used it as a model for his post-Darwinian theory of natural selection. Fisher conceived the object of evolutionary theory to be *not* a population of individual organisms, as Darwin had done, but a massive ensemble of indefinitely many abstract entities. These he called 'gene types' or 'gene ratios'. 'It is often convenient to consider a natural population not so much as an aggregate of living individuals but as an aggregate of gene ratios' (Fisher 1930: 34).

Fisher assumed that the average contribution of each gene type to the growth rate of the population is miniscule. Further, he assumed that the effect of each type on the dynamics of the population is independent of the effects of any other trait type. These assumptions allowed him to represent evolution as a phenomenon that occurred to vast ensembles of

---

[11] The definitive source on these debates is Provine (1971).

abstract gene types, in just the way that increase in entropy occurs to vast ensembles of molecules in a gas.

Response to selection, he argued, could be analysed through a kind of statistical mechanics: instead of following the effects of individual genes on a selected phenotype, one could calculate their aggregate effects under the assumption that a character is underlaid by an infinite number of genes, each unlinked to all others . . . . (Orr 2005: 120)

Given these quite radical abstractions and idealisations, Fisher could describe the way that a population undergoing natural selection changes its structure. Each trait type is assigned a fitness, an 'intrinsic rate of increase'. Fisher supposed that a population that varied in the intrinsic rate of increase of its gene types would manifest two intimately related kinds of changes. As faster growing gene types increase in their relative frequency at the expense of slower growing gene types, the population growth rate increases, and the variation in growth rates diminishes. With this in mind, he advances his Fundamental Theorem:

The rate of increase of fitness of any [population] at any time is equal to the additive genetic variance at that time. (Fisher 1930: 36)

In articulating the Fundamental Theorem, Fisher is explicitly framing an analogy to the second law of thermodynamics. He immediately draws our attention to their commonalities:

It will be noticed that the fundamental theorem . . . bears some remarkable resemblances to the second law of thermodynamics. Both are properties of populations, or aggregates, true irrespective of the nature of the units which compose them; both are statistical laws; each requires the constant increase in a measurable quantity, in the one case the entropy of the physical system and in the other the fitness . . . of a biological population. (Fisher 1930: 36)

This is population thinking in its most rarefied form. It involves 'ignoring individuals' in a comprehensive and radical way. Individual organisms are no part of the ontology of Fisher's theory, nor do they figure in its explanatory apparatus.

The whole investigation may be compared to the analytic treatment of the Theory of Gases, in which it is possible to make the most varied assumptions as to the accidental circumstances, and even the essential nature of the individual molecules, and yet to develop the general laws as to the behaviour of gases. (Fisher 1930: 36)

Margaret Morrison vividly captures the distinctively abstract and statistical nature of Fisher's thinking.

Essentially he treated large numbers of genes in a way similar to the treatment of large numbers of molecules and atoms in statistical mechanics. By making these simplifying and idealizing assumptions, Fisher was able to calculate statistical averages that applied to populations of genes in a way *analogous* to calculating the behavior of molecules that constitute a gas. (Morrison 2002: 58–59, emphasis in original)

It is extremely important to understand the degree to which Fisher's statistical approach to population change is a departure from Darwin's causal account of population change. In a certain sense, they aren't even about the same kind of thing. Darwin's theory plots the change in frequency of *lineages of organisms* in a population, as a function of their success in the struggle for life. Fisher's theory of natural selection plots the change in *intrinsic growth rate* of an indefinitely large population of *abstract 'gene ratios'*, as a function of its statistical distribution of growth rates. They don't appear to have a whole lot in common.

2.4.1   Interpreting the Modern Synthesis

Darwin's theory has a reassuring transparency that goes hand in hand with its metaphysical modesty. It needs no special interpretation, as it deals in the eminently biological phenomena of living, dying, reproduction and resemblance of organisms. Modern Synthesis population dynamics, however, is decidedly more abstruse. In ascending to its new ontology of gene types, the Modern Synthesis theory of natural selection introduces a panoply of theoretical entities and quantities, abstractions and idealisations: from gene ratios, to intrinsic rates of increase, from indefinitely large populations, additive genetic variances. In this respect, it is more like Einstein's General Relativity than Darwin's descent with modification. The theory of population dynamics enshrined in the Modern Synthesis provides us with a powerful, predictive and quantifiable theory of the dynamics of population change. But with this power, come challenges of interpretation. The Modern Synthesis trades transparency for potency.

Interpreting an abstract model is by no means a straightforward affair. Nancy Cartwright (2010) likens a scientific model to a parable or a fable. Fables and parables are used as heuristic and didactic tools for understanding and inculcating complex features of human social interaction. While folk tales may pack a powerful, salutary message, they can sometimes be a little obtuse. Generally, their import is lost if they are interpreted literally. Aesop's fable of the fox and the crow, for example, imparts an important lesson, but it isn't that corvids and canids can converse.[12] In like manner, abstract scientific models can be powerful tools, but there too we need to exercise care in drawing metaphysical morals from them. We cannot generally read the commitments of a scientific theory

---

[12] I think it's that one shouldn't succumb to flattery, or sing with one's mouth full, or something like that.

directly from the ontology of its models, especially models so abstract and idealised as those of Modern Synthesis population dynamics.

So constructing the model and deriving its consequences are just a small step towards drawing a lesson from it. In order to know what the parable means we need to study a great deal of text, reading both the theory that imbeds the model and reading the world itself. (Cartwright 2010: 31)

The question arises, then, how we *should* interpret the population dynamics models of the Modern Synthesis. What do they say about how evolution works? This question is relevant to the overall project of delineating the role of organisms in evolution in two separate ways. The first is the issue of what, according to the Modern Synthesis, causes evolution. The other revolves around the conditions under which the dynamics of populations, as described by these models, realises adaptive evolution.

The causes of evolution   The standard approach to interpreting Modern Synthesis models is to take them at face value. This reading supposes that, like Darwin's theory, the Modern Synthesis theory articulates the causes of population change. Selection models cite the processes of selection and drift and define them over the distribution of trait fitnesses. Selection is that change in a population predicted by the variation in trait fitnesses. Drift is that change that deviates from the outcome predicted by the fitnesses. On this interpretation, then, fitness is a causal property of a trait type, its propensity to increase or decrease in relative fitness.[13] Selection and drift are causal processes. They move a population from one state – trait distribution – to another, according to the variation in fitnesses.

Elliott Sober gives the definitive statement of the causal interpretation.

In evolutionary theory, the forces of mutation, migration, selection and drift constitute causes that propel populations through a sequence of gene frequencies. To identify the cause of a population's present state ... requires describing which evolutionary forces impinged. (Sober 1984: 141)

The sentiment is echoed by the population geneticist Graham Bell. He begins his influential textbook on evolutionary genetics by stating that:

The main purpose of evolutionary biology is to provide a rational explanation for the extraordinarily complex and intricate organization of living things. To explain means to identify a mechanism that causes evolution, and to demonstrate the consequences of its operation. (Bell 2008: 1)

---

[13] Alternatively, on Sober's (2013) recent view, trait fitness is not a causal propensity of a trait type, but variation in trait fitness is a causal disposition of a population.

For Bell, selection is that mechanism. So, on the orthodox interpretation, population models are mechanistic models that explain the change in trait structure in a population by citing the causes of that change in trait structure.

Recently, an alternative interpretation has arisen, the statistical interpretation.[14] It holds that Modern Synthesis models provide a sophisticated statistical description of the rates of change of trait types in a population. While these models permit us to predict, quantify and explain those changes, they do not explicitly represent the causes of those changes. If we want to study the causes of evolution we must look elsewhere.[15] Those who advocate the statistical interpretation tend to suppose that the *causes* of evolution are just as Darwin said they are, the several activities of individual organisms.

In effect, we might think of this dispute as a disagreement about which sorts of models – Darwinian or Fisherian – most accurately capture the causes of population change. The outcome of the debate is crucial for our purposes because if the orthodox view is correct, then one can describe the *causes* of evolution without ever having to consider organisms. It leaves no causal or explanatory work for organisms to do in the evolutionary dynamics of populations. Where there is no work for them to do, they are theoretically dispensable. It is to be expected, then, that individual organisms and their properties should be marginalised in evolutionary thinking.

Population dynamics and adaptation   One decided advantage of Darwin's approach to natural selection is that it is obviously an account of adaptation. Darwin's process demonstrates that, all things being equal, the lineages that increase in a population consequent on the struggle for survival are those that are better suited to the conditions of existence. The change in population structure is one in which better suited *individuals* replace less well-suited ones. In the process, a significant biological property – *individual adaptedness* – increases as a consequence. Explaining adaptation, 'fit', of course is one of the principal objectives of a theory of evolution. Similarly, in Modern Synthesis theory, selection leads to the increase in a particular parameter too: the intrinsic growth rate of a population. However, it isn't at all clear what the relation is between the intrinsic growth rate of a population and the adaptedness of individuals in it.

Consequently, there is a lingering question concerning how closely the phenomena represented by Modern Synthesis models map onto the biological phenomena of interest to evolutionists. One particular difficulty with Fisher's theory is that it is well known – indeed predicted by Darwin's own theory – that

[14] See Matthen and Ariew (2002); Walsh (2007a, 2010a); Walsh, Lewens and Ariew (2002).
[15] Marjorie Grene appears to have been the first to appreciate this distinction: '[we] must ... distinguish between "genetical selection," which is purely statistical, and Darwinian selection which is environment-based and causal' (1961: 30).

populations do *not* increase in their growth rate under natural selection. Darwin's fanciful example of elephant overproduction illustrates the way that selection imposes a check on the natural tendency of a population to continue its growth rate indefinitely. Yet Fisher's model posits an increase in intrinsic rate of growth of the entire population. This is clearly a counterintuitive – indeed counter-to-fact – consequence.

Fisher squares this incongruity of theory and fact, by making a distinction between the *intrinsic* growth rate of a population and its total growth rate (all things considered). He does so by distinguishing two kinds of genetic variance. On the one hand, there is additive genetic variance. Holding constant the environment, Fisher says, additive genetic variance always decreases. And as it does so, the *intrinsic* rate of growth of the population *always* increases. This tendency he called 'natural selection': '. . . the updated meaning of [Fisher's Theory of Natural Selection] is that there is an intrinsic connection between natural selection and a trend toward maximization' (Huneman 2014: 16). There is also *non* additive genetic variance. This is variance in the growth rates of gene types that is due not to their intrinsic capacities, but largely to their interactions with their environments (including other gene types). As populations change, so do the environments in which they are situated. This interaction generally causes a diminution in the growth rate of a gene type. Fisher called this phenomenon 'environmental degradation'. The total rate of change in fitness in a population is the sum of the change due additive genetic variance – 'natural selection' – and the portion of the change due to environmental degradation. This is an extremely sophisticated and elegant way of preserving the idea that natural selection invariably increases the intrinsic growth rate of a population. But it risks divorcing the Modern Synthesis theory of population dynamics from the behaviour of actual biological populations, except in those circumstances in which we artificially hold the environment constant.[16]

Given the possibility of environmental change, it seems unlikely that anything general can be said about either the magnitude or direction of the total change in average fitness, in real biological populations. But if we hold fixed the environment and consider only the partial change that results from natural selection, something general can be said, Fisher believed, namely, that this partial change is given by the additive genetic variance. (Okasha 2008: 331)

Fisher's conception of natural selection, then, is a highly contrived and artificial one. It turns out not to have much to do with the kinds of changes we observe in natural populations, and bears the merest passing resemblance to the process to which Darwin gave the same name. Therein lies the problem. It just isn't clear what natural selection, construed in this way, tells us about adaptive

---

[16] I return to this issue in Chapter 10.

evolutionary change. Indeed, as Fisher made a special point of saying, 'natural selection is not adaptation' (1930: 1). In consequence, the Modern Synthesis theory of evolution is not, obviously at least, a theory of how populations come to comprise organisms increasingly well suited to their conditions of existence, in the way that Darwin's theory manifestly is.

Then what is the relation between natural selection as construed by the Modern Synthesis population dynamics, and adaptive evolutionary change? This has by no means been a trivial question for theoretical population biology (Orr 2005, 2007a).[17] The evolutionary geneticist, Richard Lewontin evinces an uncommon sensitivity to the problem.

A description and explanation of genetic change in a population is a description and explanation of evolutionary change only insofar as we can link those genetic changes to the manifest diversity of living organisms.... To concentrate only on genetic change, without attempting to relate it to the kinds of physiological, morphogenetic, and behavioral evolution that are manifest in the fossil record and the diversity of extant organisms and communities, is to forget entirely what it is we are trying to explain in the first place. (Lewontin 1974: 23)

The salient question, then, is under what conditions do the changes described by Modern Synthesis models realise *adaptive* evolutionary change, the Darwinian process in which populations come to comprise organisms increasingly well suited to their conditions of existence.

2.4.2 Continuity and quasi-independence

Lewontin (1978) identifies two central requirements for changes in population structure to realise adaptive evolution. 'Continuity and quasi-independence are the most fundamental characteristics of evolutionary processes. Without them organisms as we know them could not exist because adaptive evolution would have been impossible' (1978: 230). Continuity requires that small differences in organisms correspond with small differences in individual fitness. Quasi-independence requires that individual character traits may change more or less independently.

One popular line of thought is that continuity and quasi-independence are vouchsafed by the intrinsic properties of gene tokens. The idea is that gene tokens are the concrete realisers of Fisher's abstract gene types. A gene token individually makes a typically miniscule causal contribution to the fitness of the individual it is in. Moreover, the contribution that a gene token makes to an individual's character is reasonably independent of the contributions made by the individual's other gene tokens. The effects of genes on individual fitness,

---

[17] Orr and Coyne (1992) discuss the assumptions made by Fisher about the process of adaptive evolution presumed in his models. They conclude that the 'neo-Darwinian view of adaptation ... is not strongly supported by the evidence' (Orr and Coyne 1992: 725).

thus, are more or less context insensitive and additive. To be sure, there are epistatic effects, but these can be hived off as the nonadditive effects of a gene's environment on form and fitness, and assumed collectively to be small. Owing to the independence and additivity of gene *tokens*, the contributions made by the growth rate of gene *types* to the growth rate of the population overall are more or less additive too. It makes sense, then, to distinguish the (additive) effect on population change due to the intrinsic properties of gene tokens, from the (nonadditive) effect of their environments, and to apportion the lion's share of evolutionary change to intrinsic properties of genes.

According to this line of thought, Fisher's simplifying assumptions of the additivity and independence of trait fitnesses work because they capture something of genuine importance about activities of gene *tokens* – their effects on phenotypes and fitness are to a considerable degree context insensitive and mutually independent.[18] Insofar as evolution is adaptive, it is due to the aggregated contributions of gene tokens to individual fitness. To the extent that evolution isn't adaptive, it is due to the interaction of genes with their environments (and with one another). On these grounds, Okasha credits Fisher with having made a powerful theoretical argument for gene-centrism, one that invites a radical form of antiorganicism.

Once we think in terms of a population of genes, rather than diploid organisms, it no longer looks so odd to argue that the selective environment changes when gene frequencies change, even in the absence of any fitness-affecting interactions between organisms. For what matter are the interactions between the genes, not the organisms. (Okasha 2008: 338–339)

Nowadays, a wholly different suggestion is being canvassed. The alternative proposes that the conditions for adaptive evolution – continuity and quasi-independence – are *not* held in place by the intrinsic properties of genes. In fact, the proposal goes, adaptive evolution would be impossible if the effects of genes on phenotypes (and on fitnesses) were reciprocally independent as the gene-centric hypothesis supposes them to be (Kauffman 1993; Wagner and Altenberg 1996). On the competing hypothesis, the conditions for adaptive evolution are secured by the self-organising dynamics of organisms, particularly as seen in their development. This dynamics, in turn, virtually guarantees that the effects of gene tokens on individual character (and on fitness) are generally neither independent nor context insensitive; nor are they additive. They are highly complex, nonlinear, cyclical and context-sensitive, in just the way that we characterised the dynamics of complex, self-organising systems in the preceding chapter. The upshot is that the role of genes in evolution cannot be understood in isolation from the influence of the whole organisms of which they are a part.

---

[18] This line of thought owes an obvious debt to the 'analytic method'.

We shall have the opportunity to explore the latter line of enquiry in more detail in later chapters. For the present it leaves us with two intriguing thoughts. The first is that the case for gene centrism in adaptive evolution may rest on a facile, overly literal interpretation of what are in fact highly abstract models of population change. The other, more important, implication is that the distinctive features of organisms may be necessary for explaining why adaptive evolution is adaptive. I raise the issue here simply to draw to attention to the fact that the practice of 'ignoring individuals' inaugurated by the Modern Synthesis version of population thinking risks the prospect of omitting an important part of the story of evolution. It potentially leaves us without a satisfactory answer to the very fundamental question 'what makes adaptive evolution adaptive?'.

Conclusion

We have, in effect, two evolutionary theories: one we inherited from Darwin, and the other codified in Modern Synthesis population dynamics, primarily, I maintain, by R.A. Fisher.[19] They are both, in their way, instances of population thinking. Yet, there are enormous differences between them, and these have largely gone underappreciated. For our purposes, the most significant difference is the role that each accords to organisms in evolution. Darwin's theory has individual organisms as part of its explanatory apparatus. Its Modern Synthesis successor does not. The Modern Synthesis version is a powerful and sophisticated theory of the dynamics of evolving populations. It allows us to quantify, predict and explain changes in the trait structure of populations in a way that Darwin's rudimentary form of population thinking was powerless to do. It does so only by citing the relative growth rates of abstract trait types.

Interpreted literally, Modern Synthesis population models seem to suggest that the activities of individual organisms are explanatorily irrelevant. So long as the realisers of the abstract trait types – gene tokens – act in a more or less independent, context insensitive way we need only to invoke genes and their environments to explain population change. Modern Synthesis population dynamics allows us to look right through the confoundingly intractable complexity of organisms. The widespread adoption of Modern Synthesis (Grade 2) population thinking may go some significant way to explaining the extirpation of organisms from evolutionary theory.

But the apparent dispensability of organisms from Modern Synthesis population dynamics may be illusory. There is more to organic evolution than change in the frequency of trait types. To lose sight of this, as Lewontin reminds us, '. . . is to forget entirely what it is we are trying to explain in the first place'

---

[19] That said, I am willing to go along with history of biology custom and extend the kudos to Wright and Haldane.

(1974: 23). What we are trying to explain in the first place is the fit and diversity of biological form. For that we need to understand why the processes modeled by Modern Synthesis population dynamics tend to realise adaptive population change. There may yet be a substantive role for individual organisms to play in the explanation of adaptive evolutionary dynamics.

3   The fractionation of evolution

Struggling or replicating?

Bolas spiders (*Mastophora sp.*) attract their prey, male moths, by producing a substance that mimics the pheromones normally emitted by virgin female moths. As the prey approaches, the bolus spider nabs it with a ball (a bolas) on the end of a filament of enormous tensile strength that it has manufactured from its own silk glands. The bolas is coated in a special adhesive ideally suited to flowing beneath and between the moth's scales and attaching to its carapace. The spider ensnares the moth and reels it in on the filament (Shapiro 2014). That's a remarkable way to make a living, but bolas spiders are by no means alone in their bizarreness. Organisms make the outlandish routine. Take *Welwitschia mirabilis*, a vascular plant that appears to be a relic from the Jurassic. It inhabits the Namib desert, among the driest and hottest places on earth, where virtually no other plant can thrive. Yet each individual can persist for up to 1,500 years. It survives by absorbing the minute concentrations of water vapour through its two enormous leaves, which may be up to nine meters in length.[1]

Just about any snippet of natural history eloquently advertises the astonishing ability of organisms to thrive in their conditions of existence. And organisms do so in a bewildering variety of ways. The principal objective of comparative biology is to explain this fit and diversity of organic form. We have a remarkably powerful yet simple theory that does the job, the theory of evolution.

As we saw in the preceding chapter there are actually two such theories, although the differences between them often go unremarked. One of them originated in the nineteenth century with Darwin's *Origin of Species*; the other was forged, expanded and entrenched throughout much of the twentieth century. They differ primarily in what each takes as its canonical unit of biological organisation. Each draws on a different kind of entity to ground its explanations of fit and diversity. In the preceding chapter we saw how the Modern Synthesis approach to population dynamics abstracts radically away from the individual-level causes of population change. It explains population

---

[1] Natural History Museum, London (2014). http://eol.org/pages/1156352/details.

change as the manifestation of differential growth rates of abstract trait types. The fundamental lesson from Darwin, though, is that howsoever we choose to *represent* population change, it is *caused* by individual-level processes. Even if we think that the population-level parameters that explain evolutionary change – say the distribution of trait fitnesses – pick out causal properties of populations (Sober 2013), it is still the case that the causal properties of populations are inherited from the causal properties of their members. The question, then, is what are the entities that fix the explanatorily relevant properties of populations. Alternatively, we might ask, what is the unit of biological organisation that most effectively represents the individual-level causes of evolutionary change. For Darwin's theory, the fundamental unit of organisation is the organism. For Modern Synthesis theory it is the gene or 'replicator'.

It is in this transition from Darwinian theory to replicator theory that we see the most compelling reason for the loss of organisms from evolutionary theorising. Organisms are not needed because the causal capacities of genes can explain everything that Darwin called upon the capacities of organisms to explain. G.C. Williams' *Adaptation and Natural Selection: A Critique of Some Current Evolutionary Thought* (1966) is the *locus classicus* of the idea that every evolutionary phenomenon that can be explained by adverting to the properties of organisms can also be – and should also be – explained by adverting to the properties of their genes. Williams' argument appeals to longevity and stability. He maintains that the entities that are subject to adaptive improvement through selection must have 'a high degree of permanence and a low rate of endogenous change . . .' (1966: 23). Genes fit the bill ideally. Organisms come and go, but genes are the enduring atoms of evolution: 'If there is an ultimate fragment it is, by definition, "the gene" . . .' (1966: 24). In order to understand the process of adaptive evolution, then, we must look to the properties of genes. The category *organism* is dispensable to evolutionary theory.

So comprehensively had organisms been lost to evolutionary thinking toward the end of the end of the twentieth century that the functional morphologists Louise Russert-Kramer and Walter Bock struck up a posse to find them. They organised a workshop in New Orleans in 1987 dedicated to the investigation of the role that organisms might play, with the forlorn title *Is the Organism Necessary?*. Their preface to the published proceedings strikes an appropriately plaintive note.

Organismic biology is as yet an under-nourished orphan in biology, unwanted because it is not understood by biologists interested in levels above and below the central one of individual organisms. (Russert-Kraemer and Bock 1989: 1060)

Indeed one philosopher in attendance was incited to ask: *Do Organisms Exist?* (Ruse 1989). That the question should not be summarily dismissed as absurd reveals much about the state of evolutionary biology at the time. The place of organisms in biology had been so comprehensively eclipsed by replicators that it had become a genuine puzzle, even to the most committed organismal biologists, just how the category of the organism might inform our understanding of evolution.

After conceding that, yes, 'obviously in one sense they do [exist]' (1989: 1061), Ruse proceeds to outline the ways in which modern science has relieved us of any obligation to deal with them directly. He concludes ('not reluctantly') that organisms are 'teleological', 'end-directed' entities. But this presents no mystery, nor any challenge to orthodox evolutionary thinking. 'It is a direct result of the force that shapes the whole of the organic world: natural selection' (Ruse 1989: 1066).

Ruse's solution is only a partial reaffirmation of organisms, however. The category is allowed to remain in our ontology, but only insofar as it does not interfere with our study of evolution. The general ethos that Ruse is giving voice to is one in which organisms are simply incidental to – indeed, accidents of – adaptive evolution. Evolution produced organisms, but it might not have done so. Their existence is not necessary for evolution to have occurred. Replicators, however, are required.

Replicators exist. That's fundamental. Phenotypic manifestations of them, . . . may be expected to function as tools to keep replicators existing. Organisms are huge and complex assemblages of such tools, assemblages shared by gangs of replicators who in principle need not to have gone around together but in fact do go around together and share a common interest in the survival and reproduction of the organism. (Dawkins 1999: 262)

Organisms are thus seen as consequences of biological evolution, rather than as preconditions. The nature of organisms plays no indispensable role in our understanding of the process of evolution. This is quite a departure from Darwin's insistence that the activities of organisms constitute the conditions necessary for biological evolution.

In the preceding chapter, we saw Darwin's theory usurped, as it were, from above. His account of population change as a consequence of individual struggle was replaced by a theory of population change as a consequence of the differential growth rates of abstract trait types. This abstract theory, elegant and powerful as it is, has its limitations. I suggested that while it is superior to Darwin's as an account of population dynamics, it is inferior to Darwin's theory as an account of the metaphysics of evolution. That leaves open the possibility that organisms are, after all, indispensable to the study of the causes of evolution. In this chapter, we see Darwin's theory of the causes of evolution subverted from below. Replicator Biology seeks to replace Darwin's organismal concept of 'struggling' as the cause of evolution with the decidedly more molecular concept of 'replicating'. Evolution, on the view that took hold during the twentieth century, is principally caused by the activities of genes or replicators. To be sure, replicator theory has its advantages, but it also imposes on evolution a distinctive set of precepts, concepts and commitments.

3.1   Struggling or replicating?

So how has the world come to contain *Pteraelidia sp.*, *Mastophora sp.*, *Welwitschia sp.* and all its other 'endless forms most beautiful and most wonderful'? Darwin poses his question early in Chapter 3 of *The Origin*.

How have all those exquisite adaptations of one part of the organisation to another part, and to the *conditions of life*, and of one distinct organic being to another being, been perfected? . . . . (1859 [1968]: 114)

Darwin's question about fit and diversity is such a far-reaching one that it would be surprising if it were to receive a simple answer. And yet it does – a single, elegant and powerful one, as we have already seen in the preceding chapter. He gestures to it in the book's title.[2] The paragraph just quoted ends:

All these results, . . ., follow inevitably from the *struggle for life*. (1859 [1968]: 115, emphasis added)

Struggling may be a simple answer, but as an amalgam of organismal processes, it is inordinately complicated. Organisms develop, they grow, nourish, organise and protect themselves and others, compete with, mate with, prey upon, mimic and help one another. They alter their own and each others' environments. They exchange matter and energy with their environments, reproduce, nurture or abandon their young, and die. All these life processes and more are packed into Darwin's notion of the struggle for life. It would be greatly to the advantage of evolutionary biology if some simpler, more precise notion could do the job. Replication seems to be tailor-made.

3.1.1   Replicating

The replicator concept is simple and powerful. Replicators are at once the units of control over biological form, in that they possess the information – a 'code script' – from which an entire organism can be built. And they are the units of inheritance, in that they are the only entities that are copied and transmitted from parent to offspring, and yet they secure the resemblance of offspring to their parents required for genuine biological evolution.

---

[2] Which, you will recall, is: '*On the Origin of Species or the Preservation of Favoured Races in the Struggle for Life*'.

What are transmitted from generation to generation are the 'instructions' specifying the molecular structures: the architectural plans of the future organism. . . . The organism thus becomes the realization of a programme described by its heredity. (Jacob 1973: 1–2)

As the Modern Synthesis developed and grew, particularly as it assimilated the insights of molecular genetics, evolution progressively came to be seen as the change in *replicator* structure in a population. It is not, as Ruse reassures us, that replicator theory *denies* the existence of organisms (Dawkins 1982). Nor does it need them.

A casual glance at the way evolutionary theory is offered for popular consumption easily confirms the ascendency of replicators over organisms.

Four thousand million years on, what was to be the fate of the ancient replicators? They did not die out, for they are the past masters of the survival arts. But do not look for them floating loose in the sea; they gave up that cavalier freedom long ago. Now they swarm in huge colonies, safe inside gigantic lumbering robots, sealed off from the outside world, communicating with it by tortuous indirect routes, manipulating it by remote control. They are in you and me; they created us, body and mind; and their preservation is the ultimate rationale for our existence. They have come a long way, those replicators. Now they go by the name of genes, and we are their survival machines. (Dawkins 1976: 19)

The rhetoric is stirring, bordering on the giddy, but even in more dispassionate moods, biologists consistently and confidently assert the eclipse of the organism from evolutionary theory.

Evolution is the external and visible manifestation of the survival of alternative replicators . . . Genes are replicators; organisms . . . are best not regarded as replicators; they are vehicles in which replicators travel about . . . . (Dawkins 1982: 82)

We are regularly encouraged, especially by proponents of replicator theory, to overlook the difference between Darwin's organism-centred theory and its successor. Indeed we are led to suppose that the latter is the natural successor to the former, that they differ only as a matter of perspective, and not on any substantive details. Dawkins attempts to assuage any lingering worries we might have about the apparent abandonment of Darwin. Replicator theory:

. . . is Darwin's theory, expressed in a way that Darwin did not choose . . . It is in fact a logical outgrowth of orthodox neo-Darwinism, but expressed in a novel image. Rather than focus on the individual organism, it takes a gene's-eye view of nature. It is a different way of seeing, not a different theory. (Dawkins 1989: iii)

These protestations notwithstanding, there are in fact quite radical differences. These will become increasingly significant as we proceed. Perhaps the simplest way to motivate the claim that these are distinct theories is to point out that Darwinian theory and replicator theory count different phenomena as instances of 'biological evolution'.

3.1.2   *Mariposa fortunatis*

There is a subspecies of butterfly (the lucky butterfly, *Mariposa fortunatis imaginensis*) that inhabits the idle reveries of philosophers of biology.[3] Let us suppose that in this (fictitious) subspecies there is no phenotypic variation in wing pattern or colour, and there is no genetic variation in the mechanism that produces wing pattern or colour. Females lay their eggs on a host plant. Eggs hatch, caterpillars feed on the host plant materials, and imprint on them. So that when the new generations of females lay their eggs, they do so on the host plant on which they hatched. Females in this population all choose the same species of host plant. Now we are asked to imagine that by some stroke of fortune, an egg happens to be dislodged from its host and falls onto a different species of plant and grows and imprints there, and develops into a viable female. It turns out that this butterfly has different wing colour and pattern, and because of the nutritional qualities of the new host plant, it is significantly more robust than its conspecifics. This female lays her eggs on her new host and her offspring are, again, distinctive in their morphology, superior in their fitness and more numerous.

We would expect that gradually the new morph would increase in relative frequency with respect to the old morph, resources permitting. The reason isn't hard to see. The new morph leaves more offspring on average than the old. We now have variation in this population in a trait that is reliably passed down from parents (mothers) to offspring. These familial variants also have fitness consequences. This variation in fitness explains changes in the structure of the population. We may also suppose that this is genuinely adaptive change in population structure. The new morph individuals (lucky lucky butterflies) are better adapted to their environment in that they are capable of reliably exploiting it in a way that enhances their survival and reproduction over the old morph (hapless lucky butterflies). The question we ask is: is this evolution by natural selection or not?

A thoroughgoing Darwinian ought to be inclined to see natural selection going on here. Over the course of generations, the population becomes progressively well adapted to its conditions of existence. It increasingly comes to exploit a resource that enhances the reproductive output of individuals. It does so as a consequence of the differential capacities of variant individuals to succeed in the struggle for existence. The variant characters contribute to this differential success. And they reliably recur generation on generation. This looks like genuine inheritance insofar as offspring resemble their parents more than they resemble arbitrarily chosen individuals from their parents' generation. Moreover, changes in the constitution of the population are predictable

---

[3] The scenario is taken from Mameli (2004).

just on the basis of fitness differences. This certainly has the character of a genuine case of Darwinian evolution.[4]

The replicator theorist ought to be drawn the other way. From the replicator perspective, this doesn't look like a genuine case of evolution at all, much less adaptive evolution, for a number of very simple reasons. The first is that there is no change in the structure of the population, at least in the sense that this matters for replicator theory. *Ex hypothesi* the distribution of gene frequencies in the population is the same generation on generation, at least as regards genes that specify the character in question. Because there is no genetic variation, there is no variation in gene fitnesses. Consequently, there is no selection. Whatever changes there are in population structure cannot be predicted by the distribution of gene fitnesses alone. Moreover, the differences between the lucky and hapless lucky butterfly lineages are not inheritable. They are 'non-genetic' and environmentally induced. They may be retained within lineages, but that, according to replicator theory, doesn't count as inheritable. Inheritable differences between lineages must correspond to differences in their respective replicators. This is not a case of evolution.

It matters little whether there really is evolution going on here, least of all because this is a fictitious example.[5] What matters for our purposes is that the *Mariposa fortunatis* scenario points to a significant difference between the process of evolution as conceived under the Darwinian theory and the process of evolution from the perspective of its replicator-based successor. They are different theories, and our objective here is to investigate the differences, and to reconstruct the process by which one transmogrified into the other. Later we will have occasion to enquire whether the theoretical differences have any implications for our study of evolution.

3.2   Fragmentation

Evolution comprises four very different sorts of processes: development, inheritance, adaptive change in population structure and the production of evolutionary novelties. Inheritance is an interorganismal process, the process by which the traits of parents are passed on to offspring. Development is a within-organism process. It is the process by which a single cell established by fertilisation becomes a complex, differentiated adult.[6] Likewise, we are told, the generation of evolutionary novelties occurs entirely within organisms. Adaptive change is a supra-organismal process, in which populations come

---

[4] It is also worth noting that this population is susceptible of treatment by the usual models of phenotypic evolution, the Breeders' Equation and the Price Equation.

[5] The eminent lepidopterist Dick Vane-Wright has drawn my attention to what might be a genuine example. See Singer and Thomas (1996).

[6] In sexually reproducing organisms, at least. Single-celled organisms can also be said to develop.

to comprise individuals that are well adapted to the conditions of their existence. These component processes together make up evolution.

3.2.1   Components of evolution

One signal difference between Darwinian theory and replicator theory is the set of commitments that each makes to the nature and relation of these component process of evolution. Darwinian theory is somewhat noncommittal, but Replicator Biology takes a definitive and compelling stand on the way these processes combine to produce evolution. Part of the appeal of Replicator Biology lies in the capacity of the replicator concept to streamline our thinking about the component processes and the relation between them.

Development   Early in the twentieth century, biologists began to speak of genes as encoding instructions, information or a program for building an organism. The expression of genes eventuates in the synthesis of proteins, the ultimate building blocks of organisms. In guiding the production of proteins, genes are said to carry the information or the blueprint for the development of the organism. Because genes alone encode phenotypic information, they are the ultimate difference makers, they make the difference between an organism having one evolutionary trait or another (Waters 2007). Replicators, then, become the units of phenotypic control. They are the entities from which the production of form proceeds, and the factors in which evolutionary differences in form reside. Indeed, under the auspices of the Modern Synthesis development has been reconceptualised as the decoding of genetic information.

Developmental biology can be seen as the study of how information in the genome is translated into adult structure .... (Maynard Smith 2000: 177)

Inheritance   Similarly, replicators stake their place at the centre of the Modern Synthesis theory of inheritance. Replicators are the units of inheritance because they determine the similarities of evolutionary significance that hold between parents and offspring. They secure the intergenerational constancy of form. So effective has the replicator theory of inheritance been that the concept of inheritance has undergone something of a redefinition. Prior to the Modern Synthesis theory of evolution, 'inheritance' stood for a particular pattern of difference and resemblance amongst organisms – *viz.* the transgenerational constancy of interlineage differences in form. It has now come to mean the transmission of replicated entities, genes, from parent to offspring (Gayon 2000).[7]

[7] Müller-Wille and Rheinberger (2012) trace the history of changes in the concept of inheritance that led to its distinctive use in the Modern Synthesis.

Adaptive change   Replicators have also become the units of evolutionary change. Evolutionary change is change in gene frequencies. Rates of change in a population are measured by replicator (gene) fitnesses. Selection is said to occur in a population only when there is variation in gene fitnesses. 'Replicator selection is the process by which some replicators survive at the expense of others' (Dawkins 1982: 82). As early as 1937 biologists – even those attuned to the importance of organisms in evolution – were equating the study of evolution with the study of changes in gene frequencies: 'Since evolution is a change in the genetic composition of populations, the mechanisms of evolution constitute problems of population genetics' (Dobzhansky 1937: 11).

Evolutionary novelties   According to replicator biology development is essentially conservative, as is inheritance, of course. They introduce no novelties of their own. Selection, for its part, merely diminishes the variety of replicators. Without the generation of novelties, evolutionary change cannot be sustained. Random mutations of genes are the ultimate source of evolutionary innovations. Since they

constitute the only possible source of modifications in the genetic text, itself the sole repository of the organism's hereditary structures, it necessarily follows that chance alone is at the source of every innovation, of all creation in the biosphere. (Monod 1971: 112)

Replicators are not only pivotal to the component processes of evolution severally, they also demonstrate how those processes combine to produce fit and diversity. Replicators copy themselves, aggregate and interact. They build organisms and deliver them to the arena of selection. Those replicators that reside in the organisms that survive the struggle for life long enough to reproduce are reintroduced into the pool of available replicators to be copied, translated into organisms and sorted again in the sieve of selection. As less well-adapted replicators are lost from a population, they are replaced by novel replicators, and combinations thereof, whose ultimate source is random genetic mutation. Replicators are thus crucial to development, inheritance, adaptive population change and the origination of evolutionary novelties, all the processes required for evolution.

3.2.2   *Quasi-independence*

The switch to replicator biology allows biologists to treat the component processes of evolution as largely independent of one another. Inheritance, for example, is independent of development and adaptive change. An organism gets from its parents just the replicators that the parents transmit. This information is not affected by the process of development. Nor is inheritance adaptively biased. Parents do not transmit to their offspring the traits that it would be good for them to have; they simply transmit the replicators they actually possess. Similarly, development exhibits a degree of autonomy from both inheritance and adaptive change. To be sure, offspring develop the traits 'encoded' in the replicators their parents transmit to them. But the process of development does not affect the inherited information. It remains immune to alteration by the downstream processes that occur in an organism. Thus replicators are preserved, to be passed on to the following generation. For its part, the process that brings about adaptive change in populations structure – selection – also exhibits a degree of independence from inheritance, development and the initiation of novelties. Selection is clearly dependent upon what traits are inherited in a population and what organisms are developed, of course. Selection operates over the results of these processes. But it does not change them; it winnows organisms, discerning between the existing range of inherited replicators and developmental processes.[8] We are told that evolutionary novelties occur initially as mutations. They are random in the sense that they are unbiased by the needs of organisms. In this sense mutations are unaffected by the processes that promote adaptation. They are also unconstrained by development and inheritance: neither process introduces mutations.

This quasi-independence of inheritance, development, adaptive change and the production of novelties is a great practical boon to biologists. It allows one, for example, to study the dynamics of adaptive population change – selection – without having to take into account the ways in which inheritance and development might contribute adaptive changes of their own. Similarly, it allows one to study the process of inheritance without having to disentangle it from the contribution of development or adaptiveness to inheritance. Likewise, biologists can study development without having to worry about whether it introduces adaptive evolutionary changes or how it influences heredity. The tractability that replicator theory confers on the study of evolution has been enormously beneficial to biology.

Yet, it is by no means a conceptual truth – if it is a truth at all – that the component processes of evolution enjoy this degree of discreteness and reciprocal independence. There are other theories of evolutionary change, of course, that don't posit this kind of relation. In Lamarck's account of evolution, for example, both inheritance and development are biased by adaptiveness. Organisms have within themselves an innate drive toward perfection. They are capable of improving themselves by adjusting their form to the demands of their local conditions of existence. These adaptive changes are passed from parent to offspring. Populations of organisms undergo adaptive change because

---

[8] To be sure, it selects *between* alternative developmental processes, but it does not alter those processes, just the distribution of them in a population.

individual organisms undergo adaptive change during their development. The processes of individual adaptive change, development and inheritance are all intimately intertwined.

There are other accounts of evolution that fall closer to our current evolutionary thinking that, in varying ways, posit the nonindependence of its component processes. Herbert Spencer's account of evolution, for example, accorded a significant role to selection as a source of adaptive evolutionary change. Nevertheless, he also acknowledged that the processes of development can contribute to adaptive changes in a population. 'For Spencer, organisms are adaptive as individuals prior to and apart from natural selection. This means that they tend to accommodate to their circumstances.' (Depew forthcoming). The stable distribution of adaptive forms, and their changes over time are jointly due to the supra-organismal process of natural selection and intra-organismal process of development.

Ernst von Haeckel, like Spencer an ardent Darwinian, believed that heritable evolutionary novelties arise largely through the process of development. An organism inherits its entire suite of developmental processes. Occasionally, an organism will add a stage, typically a further degree of complexity at the end of development. This additional stage is passed on to its offspring. Evolution, according to Haeckel, exhibits progressive complexification because it consists primarily in the addition and subsequent inheritance of new adaptively advantageous developmental stages. Haeckel's famous slogan 'ontogeny recapitulates phylogeny' expresses the idea that the history of evolution is made evident through the similarities and differences in developmental sequences of related organisms. Here again, is an evolutionary theory in which inheritance, development, the generation of novelty and adaptive change are all interrelated.

History hasn't treated Lamarck's, Spencer's or Haeckel's versions of evolutionary theory kindly. They fail, if they do, on empirical grounds. But they are not incoherent. The point here is simply to demonstrate that the fragmentation and quasi-independence of evolutionary processes ushered in by the Modern Synthesis is not a conceptual requirement of a theory of evolution.

We should bear in mind that Darwin's own mature view on evolution held that these processes are all inextricably linked. Darwin's inheritance theory, pangenesis, hypothesised that inheritance is mediated by tiny bodies he called 'gemmules'. These gemmules were transmitted from parent to offspring at conception. They are assembled in the germline from the various cells of an organism's body. Changes in form that arose during development could be fed back into the inherited material through the gemmules. So the generation of evolutionary novelties, and the adaptive bias in form and the inheritance of characters, in Darwin's own theory, are all bound up in the process of individual organismal development.

The simple point is that there is nothing about the phenomenon of evolution that requires the discreteness and quasi-independence of its constituent processes – inheritance, development, adaptive change and the generation of novelties. That commitment is an article of Replicator Biology dogma. It may be one of the features that commends the Modern Synthesis replicator conception of evolution. It may also be true. The significance, for our purposes, is that fractionation is a part of the theoretical machinery of the Modern Synthesis, and not a conceptual truth about evolution *per se*. Given that, a number of questions arise, for example (1) How did this fractionation occur during the growth of the Modern Synthesis? and (2) Is fractionation a prerequisite for any viable theory of evolution that is consistent with what we currently know about biological processes? The first question will occupy us for the rest of the chapter. The second question is addressed at length in Part Two.

3.3   Three Weismenn of the Synthesis

The fragmentation of evolution seems to have occurred prior to the inception of replicator biology, and appears to have developed in progressive stages. We can trace its progress through the works of three nineteenth-century biologists Darwin, Gregor Mendel and August Weismann. The result is a fractionation of evolutionary processes that is ready made for twentieth-century Modern Synthesis biology.

3.3.1   Darwin

Darwin's radical shift toward population thinking served to relocate the problematic of fit and diversity. His question, as we saw in the preceding chapter, was not 'How do individual organisms come to be well adapted to their conditions of existence?' Rather it was 'How do populations come to comprise organisms so well adapted to their conditions of existence?' Adaptive evolutionary change became change in population structure, rather than fundamentally a change in individual form. So the processes that introduce an adaptive bias into evolution must be sought in the structure of populations. Darwin's theory allows a crucial shift in the meaning of 'adaptation'. It allows us to think of a trait as *an adaptation*. An adaptation, in turn, is no longer an accommodation made by an individual organism. It is a suborganismal entity, a trait whose prevalence is to be explained by the population-level process of natural selection. Darwin may well have transformed the concept of adaptation, but his account is more or less noncommittal on the relation between adaptation, inheritance, development and the generation of novelties.

3.3.2 Mendel

Darwin didn't know how inheritance was mediated, but it posed a problem for him. Inheritance appears to be 'blending', in the sense that for many quantitative traits, at least, offspring seem to possess a value intermediate between the values of their parents' traits. The offspring of one tall and one short parent will tend to be middling in height. But blending inheritance is ill suited to Darwin's requirements. If inheritance is blending, in the way that Darwin supposed, new, rare variants would be washed out of the population, progressively attenuated as they pass from parent to offspring to grand-offspring. The rate at which blending tends to wash out an extreme variant would exceed the rate at which the variant tends to spread through natural selection.[9] The rediscovery of Mendel's work by Hugo de Vries, Carl Correns and William Bateson in 1900 appeared to solve the problem.[10]

Mendel's (1866) discoveries open up the possibility that inheritance is particulate and atomic. That is to say, what is passed on from parent to offspring is a particle or factor, that is unaffected by its subsequent development in the offspring. Such a factor can be retained in a lineage unaltered generation after generation. Its traces can go undetected for generations, only to reappear. Insofar as the offspring is a 'blend' of parents' characters, that blending takes place during development. Mendelism thus introduces the possibility of a separation between the processes of inheritance and development. It also issues in a reconceptualisation of inheritance. It is no longer a gross pattern of resemblance and difference amongst organisms; it is now the process of the transmission of factors.

Nothing about the Mendelian process of inheritance, however, suggests a separation of inheritance from the source of adaptive bias, or the origin of novelties in evolution. For all Mendel knew, these factors of inheritance may have originated in the processes within organisms that promote adaptations. In which case we would still not have acquired the Modern Synthesis separation of inheritance, adaptive change, development and the origination of evolutionary novelties. The doctrine articulated by August Weismann seems to have provided the crowning touch.

3.3.3 Weismann

Weismann strongly opposed the Lamarckian theory of inheritance. He insisted that novel characters whose acquisition is attributable exclusively

---

[9] The argument is credited to Fleeming Jenkin, and can also be found in Galton. I thank André Ariew for help here.
[10] Although it took many years and a great deal of ingenuity to demonstrate the consistency of Darwin's theory and Mendelian inheritance.

to the process of development are not passed on to subsequent generations. Alongside this stricture about what could be inherited, he proposed an account of why developmentally acquired traits *could not* be transmitted. Weismann noted that early in the development of metazoan embryos – in the blastula stage – an asymmetry emerges. There is a rapidly dividing animal pole, and a slowly dividing vegetal pole. The vegetal pole becomes the germ plasm, comprising those cells from which gametes are formed. The animal pole becomes the 'somatoplasm', which develops into the body of the organism.

Once separated, the germ plasm and the somatoplasm coexist in a sort of splendid mutual isolation. Changes wrought in the germ plasm in an individual do not redound to the developing somatoplasm, and, crucially, changes incurred by the somatoplasm during development do not filter down to the germ plasm – this is the so-called 'sequestration of the germ line'. Weismann surmised that it is because of this sequestration that there is no conceivable mechanism of Lamarckian inheritance. The entire contents of material passed on from parent to offspring in inheritance are contained within the germ plasm. The intergenerational constancy of form is secured by the continuity of the germline. Inheritance is thus exhausted by the transmission of germline material.[11]

The Weismann doctrine sets the stage for the complete sundering of the component evolutionary processes.[12] Inheritance is isolated from development. Conversely, development does not affect inheritance. Similarly, the only process that introduces adaptive bias, selection, does not alter the structure of germline cells. Any adaptive accommodation that might occur within an organism leaves the germline unaffected. Development thus does not effect evolutionary changes in population structure. So, the only conceivable source of evolutionary novelties must arise in the germline.[13]

It has escaped no one's notice that the Weismann doctrine applies only to metazoans, whereas the fractionation of evolutionary processes is thought to apply to evolution in any (Earthly) domain. However, there is an alleged principle of molecular biology that appears to entail a generalisation of Weismann's fractionation of evolutionary processes: the Central Dogma of Molecular Biology coined by Frances Crick (1958).[14]

---

[11] Weismann's concept of the continuity of the germline seems to have been almost inevitable, having been anticipated a number of times, including in Galton's 'stirp' theory of inheritance and Nägeli's concept of the 'idioplasm' (Müller-Wille and Rheinberger 2012).

[12] Or at least through the reception of his work. See Griesemer and Wimsatt (1989) and Winther (2001) for the distinction between Weismann's own views and 'Weismannism'.

[13] This by itself is insufficient to entail the Modern Synthesis conviction that these changes are random. That requires a special argument that we address in Chapter 9.

[14] John Maynard Smith (1969, 1989) explicitly equates the Central Dogma with the Weismann doctrine.

The central dogma of molecular biology deals with the detailed residue-by-residue transfer of sequential information. It states that such information cannot be transferred back from protein to either protein or nucleic acid. (Crick 1970: 561)

On the face of it, the Central Dogma doesn't appear to be about the relation between inheritance and development, much less about sequestration. Nevertheless, it suggests the right sort of asymmetry; while DNA structure effects protein synthesis, the converse does not hold. Protein synthesis does not effect changes in DNA structure. Taking DNA structure to be the material of inheritance, and protein synthesis to be the process of development, then yields something akin to the Weismann doctrine, *without* requiring germline sequestration: the material of inheritance is unaffected by the processes of development.

Even at its inception, the perceived advantage of the Central Dogma was that it underscored the distinction between inheritance and development. It does even more, it gives us separation of genotype and phenotype that is so central to the Modern Synthesis.

Such a distinction made it possible to go beneath inheritance as the mere morphological similarity of parent and offspring and investigate the behaviour, and eventually the nature, of the underlying factors (genes) that were transmitted from one generation to the next. (Roll-Hansen 2009: 457)

It further establishes the primacy of genes as the stuff of inheritance, the source of novelties, and the sole units of evolutionary change.

If the central dogma is true, and if it is also true that nucleic acids are the only means whereby information is transmitted between generations, this has crucial implications for evolution. It would imply that all evolutionary novelty requires changes in nucleic acids, and that these changes – mutations – are essentially accidental and nonadaptive in nature. Changes elsewhere – in the egg cytoplasm, in materials transmitted through the placenta, in the mother's milk – might alter the development of the child, but unless the changes were in nucleic acids, they would have no long-term evolutionary effects. (Maynard Smith 1998: 10)

With this insight, the complete fragmentation of the component processes of evolution – inheritance, development, adaptive change and evolutionary novelty – so central to Modern Synthesis replicator theory is achieved. Perhaps the most significant consequence of this fractionation has been the marginalisation of the processes that go on *within* organisms – development – from evolutionary theory.[15] The project of developmental biology becomes wholly extraneous to that of evolutionary biology.

---

[15] 'Development' broadly construed here stands for the complete suite of processes that occur within an individual organism throughout its life.

After the publication of Darwin's *Origin of Species*, but before the general acceptance of Weismann's views, problems of evolution and development were inexplicably bound up with one another. One consequence of Weismann's separation of the germline and the soma was to make it possible to understand genetics, and hence evolution, without understanding development. (Maynard Smith 1982: 6)

It is worth noting how far this conception of evolution has wandered from that originally articulated by Darwin. Darwin's theory, recall, located the cause of evolution, the struggle for life, in the processes that occur within, to and between organisms in their lifetimes. These are the very processes deemed to be of minor (if any) relevance to Modern Synthesis evolutionary theory. The Modern Synthesis view may be a 'logical outgrowth' of Darwinism, as Dawkins suggests, but it differs from its predecessor in matters of substance, and not merely as a matter of perspective.

Modern Synthesis replicator theory has altered our conception of the process of evolution so profoundly that it has even changed our common understanding of the *subject matter* of evolutionary biology. Evolutionary biology is now the study of the dynamics of populations of information-encoding, suborganismal entities: replicators. We may not have noticed. Yet so patent has it become that evolution is a phenomenon of genes that now we can observe replicator dynamics even in something so mundane as a willow shedding its catkins.

It is raining DNA outside ... Up and down the canal, the water is white with cottony flecks ... The cotton wool is mostly made of cellulose, and it dwarfs the tiny capsule of that contains the DNA, the genetic information. The DNA content must be a small proportion of the total, so why did I say it was raining DNA rather than cellulose? The answer is that it is the DNA that matters ... The whole performance, cotton wool, catkins, tree and all, is in aid of one thing and one thing only, the spreading of DNA around the countryside. Not just any DNA, but DNA whose coded characters spell out specific instructions for building willow trees that will shed a new generation of downy seeds. Those fluffy specks are literally spreading instructions for making themselves. They are there because their ancestors succeeded in doing the same. It is raining instructions out there; it's raining tree-growing, fluff-spreading algorithms. That is not a metaphor, it is the plain truth. (Dawkins 1986:111)

Plain truth it may be, but 'raining DNA' is a far cry from the 'struggle for life'.

3.4   Two spaces and a barrier

The fragmentation of evolutionary processes has given us the image of evolution occurring in two discrete spaces – genotype space and phenotype space (Lewontin 1974) with a barrier, the 'Weismann barrier', between them. The famous Weismann diagram entrenches the idea.

The diagram depicts the generation on generation constancy of form (now called 'phenotype') as an amalgam of two discrete causal processes, Mendelian

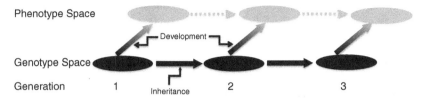

Figure 3.1 A version of the 'Weismann diagram'. Inheritance takes place exclusively in genotype space. Development maps genotype space onto phenotype space. Phenotypic 'inheritance' the resemblance of offspring to parent is not inheritance as such. It is the amalgam of two discrete processes: inheritance and development.

inheritance and development.[16] It brings into relief the Modern Synthesis redefinition of inheritance as the passing on of replicators (genes). Further, it depicts the way that development in one generation makes no contribution to inheritance in the next.

3.4.1   The dynamics of genotype space and phenotype space

On the Modern Synthesis view, *evolutionary change is change that is registered in genotype space*. The objective of evolutionary theory is to describe and explain the dynamics of genotype space.

Some processes bring about changes in genotype space directly. These include the 'genetic' processes of mutation, recombination, replication and transmission. Some processes bring about changes in genotype space indirectly, by bringing about changes in phenotype space. Differential survival and reproduction of individuals does this. When individuals die, their genes leave the population. When individuals reproduce, their genes increase in frequency in a population. So these changes in phenotype space are mirrored as changes in genotype space. Some of these changes are unpredictable, given the properties of individuals or their traits. These are sometimes called 'drift'. Other changes can be predicted on the basis of the differential capacities of individual organisms. These changes are sometimes called 'selection'. Crucially, on this view, selection is a process that causes changes in genotype space *by* effecting changes in phenotype space.

In order for selection or drift to eventuate in persistent changes in genotype space, there must be a process that in each generation maps the changes in genotype space back onto phenotype space. The distribution of genotypes in a

---

[16] Elliott Sober (1987) calls the pattern of phenotypic resemblance a 'pseudoprocess', the non-causal amalgam of two causal processes.

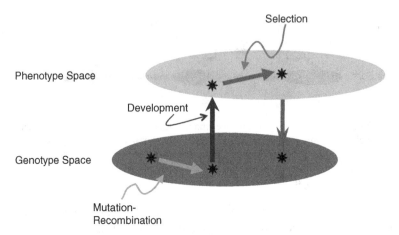

Figure 3.2 Evolution in genotype and phenotype space.
(Adapted from Lewontin 1974)

population must be translated into the distribution of phenotypes of individuals. This is the process of development. Development, as the Weismann
diagram (Figure 3.1) shows, produces phenotypes from genotypes. The
Weismann barrier ensures that there is no road back from an individual's
phenotype to its genotypes. So, development introduces no changes of its
own to genotype space.

3.4.2   The influence of development

One of the most significant consequences of the shift in evolutionary theory
from Darwin's theory of 'phenotype space' to the Modern Synthesis theory of
genotype space has been the diminished influence of organismal development
on the investigation of fit and diversity of organic form. Embryology was the
driving force behind much of preevolutionary thinking about form, but since
the advent of replicator theory it has become somewhat peripheral, unimportant
for evolutionary biology.

To some biologists, the exclusion of development has been seen as a
positive thing. Development is inordinately complex and difficult to
study. If it can be bracketed off from the study of evolutionary dynamics,
so much the better. To others, the emphasis on the dynamics of genotype
space is a matter of mere expedience: 'much of the past and the present
problems of population genetics can be understood only as an attempt to
finesse the unsolved problem of an adequate description of development'
(Lewontin 1974: 7). Indeed the burgeoning understanding of development that is emerging from recent developmental biology increasingly

appears to suggest that development has some constructive role to play in evolution.

There is a strong *prima facie* case for development playing a role in evolutionary change. A few examples might serve to illustrate the claim.[17]

Baldwin effect The American psychologist James Mark Baldwin proposed that characteristics that are initially learned socially might eventually become inherited.[18]

Animals may be kept alive let us say in a given environment by social cooperation only; these transmit this social type of variation to posterity; thus social adaptation sets the direction of physical phylogeny and physical heredity is determined in part by this factor. (Baldwin 1896: 553)

Baldwin conjectured that if a fitness-enhancing complex behaviour had to be learned, then its development in a young organism would be highly contingent on the presence of the appropriate learning cues, and the ability of the individual to pick them up. A biologically inherited trait, on the other hand, could be much more reliably developed, with a lower threshold for its initiation, and dependent on fewer specific environmental conditions (Bateson 2014).

Genetic assimilation The early developmental geneticist Conrad Hal Waddington (1956) demonstrated what seems to be a morphological analogue of this. He grew *Drosophila* embryos in a regime of extreme ether stress. Some few eggs hatched and some few of those grew into adults with two thoraxes, rather than the single thorax characteristic of dipterans. He found that after repeated crossings of bithorax individuals in increasingly less severe ether regimes, he was able to breed a lineage of bithorax *Drosophila* that required no ether shock to elicit the unusual structure. Waddington referred to the process in which these evidently acquired traits became heritable as 'genetic assimilation'.

West-Eberhard conjectures that . . .

. . . most phenotypic evolution begins with environmentally initiated phenotypic change . . . Gene frequency change follows, as a response to the developmental change. In this framework, most adaptive change is accommodation of developmental-phenotypic change. (2003: 157–158)

Consciously echoing Waddington, she calls this phenomenon 'genetic accommodation'. 'In genetic accommodation . . . phenotypes can incorporate environmental factors into normal development' (West-Eberhard 2003: 499).

---

[17] These cases receive more detailed treatment in subsequent chapters.
[18] Similar notions were proposed independently at the same time by Henry Fairfield Osborn and C. Lloyd Morgan.

Phenotypic plasticity   Organisms evince an astonishing degree of adaptive plasticity in their development. It permits them to direct their own development reliably and adaptively across a range of unpredictable and potentially lethal circumstances. This plasticity confers on organisms an ability to respond to their conditions by implementing adaptive changes to form or function of their various parts. An adaptive alteration in one subsystem of an organism, in turn, requires a compensatory change in other systems. Organisms are uniquely capable of this adaptive orchestration of their various parts and systems in development. The important point is that, as Spencer insisted, development is adaptive. Sometimes the adaptiveness of development produces evolutionary novelties. These in turn can become intergenerationally stable. In these instances evolutionary novelties appear to arise initially within development. They may later become routinised and assimilated genetically.

Prima facie examples are not hard to find (Agrawal, Laforsch and Tollrian 1999). To take a compelling putative case, Heil et al. (2004) discovered that in certain Central American *Acacia* species the secretion of extrafloral nectar, which is usually induced by environmental damage, has evolved for feeding obligately symbiotic ants. No environmental cue is needed to initiate the secretion of the nectar. The authors surmise that what in ancestral lineages had been an environmentally induced plastic response has become a robustly inherited nonplastic phenotype.

Here alterations in phenotype space, driven by the adaptiveness of development, appear to precede evolutionary changes in genotype space. Such cases may be the rule rather than the exception in evolution.

Responsive phenotypic structure is the primary source of novel phenotypes. And it matters little from a developmental point of view whether the recurrent change we call a phenotypic novelty is induced by a mutation or by a factor in the environment. (West-Eberhard 2003: 503)

Considerations such as this suggest an inversion of the primacy of genotype space over phenotype space that is so central to Modern Synthesis thinking: 'genes are probably more often followers than leaders in evolutionary change' (West-Eberhard 2005a: 6543). Increasingly for biologists, development is being seen as the principal cause of evolutionary novelty. Unlike mutation, this novelty is not random; it is biased by the fact that development responds adaptively to the needs of organisms in their conditions of existence.

These three phenomena – the Baldwin Effect, Genetic Assimilation and Phenotypic Plasticity – of course have received extensive attention. They hint, at least, that evolution might not be quite so simple as the fractionated pictures represent it as being. They suggest that the progressive marginalisation of those processes that occur within organisms that has marked out the growth of the Modern Synthesis needs to be revised, extended and in some case even replaced (Pigliucci 2009a, 2009b). Furthermore, they suggest that the component processes of evolution – inheritance, development, adaptive change and the inheritance of novelties – do not subsist in such splendid mutual isolation. At the very least these phenomena need to be 'accommodated' within the Modern Synthesis. Most of the treatments amount to attempts to reconcile these phenomena with the 'Two Spaces and a Barrier' conception of evolution.[19] Quite what implications these revisions might have for the Modern Synthesis itself is a subject for later chapters.

The fractionation of evolution appears to be a direct consequence of the shift to the replicator as the canonical unit of biological organisation. Whilst the seeds of suborganicist replicator biology may have been sown by Mendel and germinated by Weismann in earth carefully prepared by Darwin, it is far from the only thing that might have grown out of that fertile ground. In fact it seems that those who fashioned the Modern Synthesis, braiding together the various disparate disciplines of early and mid twentieth-century biology into a unified coherent whole, neither anticipated nor intended the exclusion of organisms that was to come.[20]

One of the most influential founders of the Modern Synthesis, Theodosius Dobzhansky, stressed the importance of the adaptive plasticity of organisms for the process of evolution. He certainly did not advocate the marginalisation of organisms from the study of evolution in the way that Maynard Smith suggests (Depew 2011). Dobzhansky's definition of evolution as changes in gene frequencies was intended not so much as an invitation to cast organisms aside, but as an injunction against falling prey to facile forms of Lamarckism. Ernst Mayr scorned replicator biology as 'beanbag genetics': 'To consider genes as independent units is meaningless from the physiological as well as the evolutionary viewpoint' (Mayr 1963: 263).[21] He later averred:

Evolution is not a change in gene frequencies, as is claimed so often, but the maintenance (or improvement) of adaptedness and the origin of diversity. Changes in gene frequency are a result of such evolution, not its cause. (Mayr 1997: 2092)

Perhaps, as Maynard Smith appears to suggest (above), gene-centred fractionated evolution was inspired by the successes of the molecular revolution. To understand evolution is to understand DNA. Perhaps, as Stephen Jay Gould argues (2002), it is a symptom of a sort of theoretical sclerosis that set in during

---

[19] See, for example, various contributions to Weber and Depew (2003) for discussions of the Baldwin effect, Sterelny (2009) and West-Eberhard (2003) for plasticity and genetic assimilation. In the next chapter we look at these challenges and the standard Modern Synthesis responses in more details.

[20] Joe Cain (2009) offers a compelling argument to the effect that this synthesis narrative is overplayed. The biology of the time embraces a much broader range of disciplines and viewpoints than the standard story acknowledges.

[21] J.B.S. Haldane (2008) accepts Mayr's epithet and offers up a spirited defence.

late twentieth-century biology. There the emphasis shifted to adaptation and the power of natural selection to transform the genetic structure of a population.

This is a complex story that is continually being revised and reassessed. But however it came about, the evolutionary biology of the late twentieth century was a suborganismal phenomenon comprising quasi-independent processes of inheritance, development, the origin of form and adaptive population change. Each process had its own theory; and in each theory the canonical unit of biological organisation was the replicator or gene.

Conclusion

Modern Synthesis theory gives us a fractionated picture of evolution, in which the component processes – inheritance, development, biased population change and the generation of novelty – are discrete and quasi-independent. Each, in turn, is assigned its own distinct, proprietary cause. Inheritance is the transmission of replicated genes. Development is the implementation of a program encoded in genes. Adaptively biased change in the gene structure of a population is brought about by selection. Random genetic mutation is the sole ultimate source of novelty. This division of causal and explanatory labour is extremely elegant. It serves to elevate the gene as the canonical unit of theoretical significance in evolution. Genes are crucially implicated in each of the distinct component processes severally. Moreover, genes unite these disparate processes into a single process of evolution. Evolution is what happens when genes replicate, build organisms, leave varying numbers of copies and mutate.

The fractionation of evolution in this manner is not a prerequisite of any theory of evolution. It is a substantive commitment of the Modern Synthesis. It amounts, in effect, to a bold empirical wager. The increased understanding of the processes of development and their place in evolution are now putting that wager to the test. In the next four chapters, we will enumerate some of the empirical (and conceptual) challenges that fractionated evolution faces.

Part II

Beyond replicator biology

The merits of gene-centrism are manifold. Fractionation, in particular, allows biologists to study the component processes of evolution almost in isolation of one another. As we witnessed Maynard Smith saying, thanks to the separation of development from adaptive population change, we can study the dynamics of populations without having to concern ourselves overly with the complications of ontogeny. Likewise, the mechanisms of inheritance can be studied without any particular regard to development, or the processes that bias form. The mechanisms by which evolutionary novelties are introduced into a population are, similarly, amenable to intensive study quite independently of organismal development, or the sources of population change. In turn, the effect of selection on a population can be studied without too much consideration being offered to the ways in which inheritance, development and mutation might impinge upon population dynamics.[1] It is a great practical advantage to be able to study the several component processes of evolution in relative isolation of one another. The investigation of biological processes is hard enough without having to study them all at once. But expedience is one thing, and veracity quite another.

Fractionation partially explains, even if it doesn't justify, the alienation of development in the orthodox Modern Synthesis. The division of evolutionary labour inaugurated by gene-centrism, seems to suggest that ontogeny is perhaps the least crucial component process in evolution. It is merely the process by which the phenotypic information encoded in genomes is delivered to the arena of selection, whereupon selection causes population change. To be sure it is necessary, but as long as development performs its function with reasonable fidelity, and a minimum of obtrusion into the workings of selection, its influence on evolutionary change is negligible. Small wonder, then, that development should have become the poor cousin of the evolutionary family

---

[1] In quantitative genetics, and in applications of the Breeders' equation and the Price equation, the degree of *phenotypic* heritability must be factored into the understanding of population dynamics. Here too, typically, it is assumed that the 'heritability' component of population change can be dissociated from the 'selection' component. (See Helanterä and Üller 2010; Kerr and Godfrey-Smith 2009; Okasha 2006.)

throughout the growth of Modern Synthesis evolutionary biology. And yet, as the next four chapters attempt to argue, the neglect of development may just be the principal defect of the Modern Synthesis.

Any successful theory, like Modern Synthesis evolutionary biology, does more than just gather phenomena under its ken. It also refines and revises its concepts. It is the prerogative of a successful theory to mould pretheoretic concepts to meet its needs. The Modern Synthesis has certainly done this to great effect. In addition to fractionation, it has given us a wholesale reconceptualisation of the component processes of evolution. This conceptual rejigging often goes unnoticed, or at least uncontested, but it is nevertheless of signal importance. Most significant, perhaps, is the treatment the Modern Synthesis gives to the concept of inheritance. As it appears in Darwin's theory, inheritance is a gross *pattern* of resemblance and difference. In the Modern Synthesis it has become a *process*, the copying and transmission of replicators. For its part, development too has been transformed. In the Modern Synthesis, it is not growth, differentiation and complexification as it was for the embryologists of the nineteenth and early twentieth century; it is now the implementation of a genetic program. Indeed, the entire concept of evolution itself has undergone a significant reorientation. In Darwin's theory, it is simply descent with modification; in the Modern Synthesis it is change in the gene structure of a population.

Modern Synthesis theory is certainly adequate to the challenge of explaining the processes of evolution, under its own proprietary construal of those processes. It is a quite different, and open, question whether gene-centred theory captures all that is important about the process of evolution that Darwin discovered. It is this question that motivates this part of the book.

Chapter 4 addresses the issue of inheritance. It seems evident that the *pattern* of resemblance and difference that constitutes the pretheoretical notion of inheritance might not be fully captured by the Modern Synthesis process of inheritance. Indeed, dissatisfaction with the Modern Synthesis theory of inheritance is by now generally widespread (Jablonka and Lamb 2010; Lamm 2014). The mismatch between the demands of our pre-Modern Synthesis concept of inheritance and what the Modern Synthesis delivers raises the opportunity to ask whether its revisionary account of inheritance is really fit for purpose. The most glaring defect of the Modern Synthesis approach to inheritance is precisely that it makes no provision for the various ways in which organismal development, broadly construed, can contribute to the pattern of resemblance and difference that constitutes inheritance. Organisms, as purposive, adaptive agents, actively participate in the maintenance of this pattern. Their omission has left us with a distorted and devitalised conception of inheritance.

The demotion of development as an evolutionary process is in part due to the ordination of genes as the units of phenotypic control. The genome, it is said, encodes information, or a program, for building an organism. In Chapter 5 we broach the question of what it might mean for genes to carry such information. It has often been pointed out that there is no clear way to make the 'genes as information' trope anything other than an idle metaphor. Genes may embody information, but that by no means entails that they exert any privileged control over the phenotype. I argue that there is a way to put some flesh on the bones of the 'genetic program' metaphor. But when the metaphor is spelled out we find it generally more plausible to suppose that *if* genes encode programs, it is because organisms program their genes. The gene-centred 'genetic program' approach to development founders for the very reason that its counterpart conception of inheritance fails. It neglects the active role that organisms as adaptive agents play in the control and production of their own form.

As inheritance and development introduce no systematic adaptive bias, the Modern Synthesis must designate some other process to do so. That role falls upon selection. Fractionated evolution casts selection as a wholly discrete process, independent of inheritance and development. As we discussed in Chapter 2, it is not entirely clear that this is a coherent account of the metaphysics of selection. If selection is wholly constituted of the activities of individual organisms, then the source of the systematic bias must reside within organisms. Chapter 6 argues that development is the source of the adaptive bias in evolutionary change. Evolution is adaptive because organismal development is adaptive. This is entirely antithetical to the marginal place to which gene-centred evolutionary theory relegates organisms.

Clearly, then, there are reasons to re-evaluate the neglected process of development. Chapter 7 surveys various strategies for integrating development into evolutionary thinking. It distinguishes 'Three Grades of Ontogenetic Commitment' – three levels of increasing integration of development into evolution. The chapter argues that only the third grade is adequate. In this grade, the adaptive purposiveness of organisms as manifested in their development is the source of both the pattern of inheritance and the adaptive bias in evolution.

This part of the book, in effect, constitutes a plea for a wholesale re-evaluation of the dogmas of Modern Synthesis gene-centred evolution. It argues that Modern Synthesis thinking is inadequate precisely because it fails to accord any substantive role to the purposiveness of organisms in evolution. While this section may serve to motivate the search for an alternative, it doesn't offer one. The project of articulating, or at least adumbrating, a positive alternative is postponed until Part III.

4   Inheritance

Transmission or resemblance?

If evolution is to be cumulative and persistent, then the changes wrought to the structure of a population at one time must be preserved to some degree into the next. In the process that Darwin discovered, the requisite persistence is secured by inheritance.[1] Inheritance, in this sense, is a gross pattern of cross-generational resemblances and differences amongst organisms. At the time that Darwin was formulating his theory, inheritance wasn't generally taken to constitute a single, unified biological phenomenon. Rather, it was a loose assemblage of phenomena, comprising *inter alia*, family resemblances, species-specific commonalities, persistent variation between lineages, the recurrence of pathologies, the loss and re-emergence of characters, the 'regression to the mean' of extreme traits within lineages (Müller-Wille and Rheinberger 2012). Inheritance ('heredity') coalesced into a unified object of study, more or less contemporaneously with the emergence of evolutionary theory. The synthesised study of inheritance came into biology as the product of a '... reification process, from "the heredity" as a metaphor to "heredity" as an explanatory biological concept that implies a particular kind of independent causation ...' (López-Beltrán 1994: 214). The impetus, for the emergence of heredity was provided by evolutionary thinking itself.

[H]eredity became a general biological problem when organisms acquired a genuine 'history'; that is, when the forms of life ceased to be fixed by assumed species boundaries. ... Hereditary variations indicated that life possessed an autonomous quality that was never compatible with the view that organisms had been designed for an everlasting environment nor with the view that they possessed a capacity for unlimited plasticity or perfectibility. (Müller-Wille and Rheinberger 2012: 75)

So, inheritance, or 'heredity', was inaugurated as a scientific discipline in large measure for the purpose of furthering the understanding of evolution.[2]

---

[1] Earnshaw-Whyte (2012) demonstrates that the kind of persistence required by the Modern Synthesis population dynamics models does not require inheritance *per se*. But Darwin's theory does.

[2] Greg Radick (2012) suggests that, in a quite dramatic change of heart about the propriety of the term 'heredity', William Bateson came to use it as a pedagogical tool around 1900, to teach undergraduates about evolution.

Moreover, the concept of inheritance seems to have changed along the way, keeping pace with the growth of evolutionary biology. Jean Gayon (2000) argues that inheritance transmuted between the 1850s and the twentieth century. It was considered a quantifiable magnitude – a degree of resemblance – by biologists like Galton. After the advent of Classical Genetics and the molecular revolution, inheritance became a 'structural' concept. The point is that the very idea of biological inheritance is beholden to its place in biological theorising, and is subject to revision as the theory in which it is embedded changes.

With that in mind, it is worth noting that the relation of inheritance to evolution has once again been the subject of debate since the turn of our own century. In the preceding chapter we touched upon the way that advances in evolutionary biology, particularly in molecular biology, had transformed the concept of inheritance. It has gone from being a gross pattern of resemblance and difference amongst organisms to being a process in which replicated materials are transmitted from parent to offspring. Lest anyone should doubt the comprehensiveness of this conceptual recasting, it is worth remarking that T.H. Morgan's Nobel Lecture of 1934 begins: 'The study of inheritance, now called genetics, . . .' (Morgan 1934: 1).

The question increasingly being asked these days is whether Replicator Biology's distinctive approach to inheritance is sufficient to capture everything that is of evolutionary significance about the pattern of intergenerational stability of form. It is becoming apparent to many observers that it is not: '. . . it is quite obvious that there is more to inheritance than transgenerational transmission of genetic material . . .' (Helenterä and Üller 2010: 2). Biologists are increasingly pointing to the need for a more 'inclusive' conception of inheritance (Danchin and Pocheville 2014). Critics argue that inheritance is not strictly a phenomenon of the transmission of DNA. Should it turn out that a comprehensive theory of inheritance needs to countenance more than the transmission of replicators, the repercussions would be far-reaching. 'The reason is that the existence of robust mechanisms of trans-generational inheritance independent of DNA sequences runs strongly counter to the spirit of the Modern Synthesis' (Noble 2013: 1).

4.1   From stability to replication

Darwin had a problem with inheritance. He knew that for differential survival and reproduction to lead to predictable, sustained population change, variant characters must be reliably reproduced from parent to offspring. It turns out, though, that not just any way of sustaining such resemblances will do. Darwin was well aware that in populations of sexually reproducing individuals, parents vary one from the other, and for many of the traits in which the parents vary, the offsprings' traits are more likely to resemble the mid-point of the parents' traits.

In other cases the offspring will resemble one parent rather than the other, and the trait may disappear temporarily from a lineage, only to reappear in subsequent generations. So inheritance manifests a strange mix of 'blending' and 'eliminating', and at the same time conserving. He puzzles:

> The laws governing inheritance are quite unknown; no one can say why the same peculiarity in different individuals of the same species, and in individuals of different species, is sometimes inherited and sometimes not so, why the child often reverts in certain characters to its grandfather or grandmother .... (Darwin 1859 [1968]: 13)

The tendency for 'blending' is a particular problem for Darwin's theory of evolution, especially for the evolution of novel characters. Generally, when novel characters appear in a population they are rare. Even when these novelties are advantageous in the struggle for life, they are at considerable risk of being 'blended' or 'eliminated' from the population at a greater rate than they are promoted by natural selection. The challenge was formulated most vividly by Fleeming Jenkin.

> We have abnormal variations called sports, which may be supposed to introduce new organs or habits in rare individuals ... we may suppose the offspring of the sports to be intermediate between their ancestor and the original tribe. In this case the sport will be swamped by numbers, and after a few generations its peculiarity will be obliterated. (Jenkin 1867: 294)

This is by no means a trivial problem. It is one that Darwin never solved.[3]

The year before Jenkin articulated the challenge of inheritance for Darwin's theory, however, the seeds of its solution were sown. Gregor Mendel interpreted his breeding experiments as demonstrating that in sexually reproducing organisms, for each trait, or at least for each of those he investigated in his selective breeding experiments, each offspring receives one particle, or factor, from each parent, and that in the population as a whole, each factor sorts more or less independently of all others. These factors underwrite the inheritance of characters, and are passed on unchanged from generation to generation. Factors combine to produce offspring character. Perhaps inheritance couldn't be both retentive and blending, but according to the Mendelian scheme, it doesn't need to be. Inheritance is retentive – it consists in the transmission of factors. Development is 'blending' – it consists in the joint expression of parental factors.

Thus, the particulate theory of inheritance gives us *both* the persistence of inheritable material across multiple generations, and the blending of inheritance within one generation. At a stroke Mendel had fixed Darwin's problem.[4]

---

[3] Fisher (1930) addresses the same issue in his brilliant argument to the effect that if Darwinian selection is to be a significant factor in evolution, then inheritance must be particulate.

[4] It is interesting, however, that early in the twentieth century it was thought that Mendel's discovery of factors was incompatible with Darwin's theory. Provine (1971) beautifully narrates the story of the enmity between neo-Darwinism and Mendelism, and their eventual *détente*.

This is to be sure a brilliant scientific discovery. But its history is a curious one. It lay fallow, for its first quarter century, all but ignored. It is thought that Darwin was aware of Mendel's discovery – as were many of Darwin's contemporaries – but did not see its relevance (Galton 2009). The second quarter century of Mendel's discovery ushered in a dramatic change of fortune. The rediscovery of Mendel's work by Hugo de Vries and others at the turn of the nineteenth/twentieth century instigated the conceptual change from inheritance as pattern of resemblance to inheritance construed as the transmission of replicators.

In the early twentieth-century study of inheritance, Mendelian factors, or genes, were not usually identified as discrete units with distinctive causal powers. More often than not, they were treated instrumentally.

There is not consensus of opinion amongst geneticists as to what genes are – whether they are real or purely fictitious – because at the level at which genetic experiments lie, it does not make the slightest difference whether the gene is a hypothetical unit, or whether the gene is a material particle. (Morgan 1934: 315)[5]

It was only with the advent techniques in molecular biology that genes become independent, discrete units that exerted highly specific causal control over phenotype (Griffiths and Stotz 2007).

Molecular biology was born when geneticists, no longer satisfied with a quasi-abstract view of the role of genes, focused on the problem of the nature of genes and their mechanism of action. (Morange 1998: 2)

In the process, the concept of inheritance changes. It goes from a multifarious, inchoate and poorly defined pretheoretic notion of general resemblance to a rigorously defined, intricately understood, unified, discrete and independent molecular process. The theory of inheritance that grew out of Mendel's work, and the chromosome theory of inheritance, and subsequently molecular biology, is a prime instance of what Matteo Mameli (2005) calls a 'donation/ conception' account of inheritance. In this version, what is inherited are not traits, but molecular units of phenotypic control donated to the offspring by the parent at conception. Inheritance is the transmission of replicated materials.

4.2   Acquired and inherited characters

The Weismann barrier enshrines the Modern Synthesis distinction between those traits that are objects of genuinely evolutionary change and those that are not. Real evolutionary changes are caused by the differential expression of *inherited* traits. Modern Synthesis thinking draws a sharp distinction between

---

[5] I first encountered this quotation in Falk (1986).

inherited and acquired characters, and has elevated one to a position of theoretical significance and in the process demoted the other. Inherited traits are the only traits that genuinely evolve.

4.2.1 Soft versus hard inheritance

In support of this crucial Modern Synthesis distinction, Ernst Mayr differentiates between what he calls 'soft' and 'hard' inheritance (Richards 2006). In soft inheritance,

the genetic basis of characters could be modified either by direct induction by the environment, or by use and disuse, or by an intrinsic failure of constancy, and that this modified genotype was then transmitted to the next generation. (Mayr 1982: 4)

By contrast, in hard inheritance characters modified through direct induction by the environment or by 'use and disuse' cannot affect the genetic basis of characters, and thus cannot be inherited. Most biological thinkers prior to the twentieth century, Darwin included, thought that some version of soft inheritance was a genuine possibility. That is to say that use and disuse, or environmental induction, could affect the medium of inheritance donated by the parents at conception. But the Weismann barrier renders the very idea of soft inheritance incoherent. If inheritance is the transmission of genes, and no information flows to genes from the process of development, then inheritance can in no way be sensitive to characters that arise during ontogeny. Modern Synthesis inheritance is hard inheritance. Mayr argues that the transition from soft inheritance to the Modern Synthesis conception of hard inheritance was crucial to the development evolutionary theory. Allan Wilkins concurs.

To have accepted inheritance of acquired characteristics would have meant in effect turning one's back on what was a major conceptual element that lay at the foundations of population genetics. It was the idea, backed up with a fair amount of experimental evidence by the end of the 1910s, that complex traits in animals and plants were underlain by the cumulative effects of multiple Mendelian factors. (Wilkins 2011: 130)

The insistence on hard inheritance certainly entrenches the distinction between inherited and acquired traits. It also appears to undermine the very plausibility of the inheritance of acquired characters. But it does so in a completely tendentious way. It assumes the Modern Synthesis dogma that inheritance is the transmission of replicants – genes – and then avers that genes are not altered by use and disuse of characters, or environmental induction (mutation aside). Modern Synthesis inheritance most assuredly is hard. The real question is whether the biological phenomenon of inheritance is exhausted by Modern Synthesis inheritance.

4.2.2   *Nongenetic origins*

The Modern Synthesis commitment to the inherited/acquired distinction has come under sustained challenge (Keller 2011b; Oyama 1985, 2000). It has long seemed to many biologists that there are multiple cases that violate the spirit, if not the letter, of the Weismann doctrine. In the preceding chapter we encountered a range of phenomena in which novelties appear to arise during the development of individual organisms, only later to become robustly inherited in a manner that seems indistinguishable from any other genuine form of biological inheritance. The Baldwin Effect, genetic assimilation, phenotypic plasticity all pose a *prima facie* challenge to the Weismann doctrine (Vane-Wright 2011).

The challenge is identified succinctly by Kim Sterelny (2009). He notices that novelties can occur within individual organisms, induced by the environment, or by phenotypic plasticity, or niche construction, or learning, for example—but these novelties are not *evolutionary* novelties. 'Such novelties have no effects on the germline [*sic*] are not inherited (2009: 94)'. Their initial appearances are mere 'ecological events'. He wonders how ecological events can become 'evolutionary events'.

The standard Modern Synthesis answer exploits the conviction that there is one genuinely evolutionary process that operates exclusively in phenotype space – natural selection. The introduction of a novel trait through the process of development, or through environmental influences, alters the selective regime in such way that it causes those traits to have greater fitness. A change in phenotype space ensues without any corresponding change in genotype space. As yet, there is no evolutionary change. If these phenotypes are retained in the population long enough, novel genes or gene combinations that are capable of reliably producing them arise. At that point there are 'genes for' the new phenotypes. The ecological event has become an evolutionary event. These new genes can then be transmitted through the normal course of genetic ('hard') inheritance.

Waddington's explanation of the bithorax results nicely exemplifies this strategy. He claims that before his intensive selective breeding regime, the development of the normal thorax condition in *Drosophila* is heavily canalised. By 'canalised' he means that the epigenetic interactions amongst the genes involved in the development of the thorax secure the proper development of the thorax across a wide range of circumstances and perturbations. Waddington pictures the normal development of a structure as occupying a valley in what he called an 'epigenetic landscape'. The development and progressive differentiation of a structure (what he called a 'chreode') is likened to a ball rolling from a ridge into a valley. As it progresses, it is channeled toward the final endpoint by the contours of the valley walls. If the developmental process should be

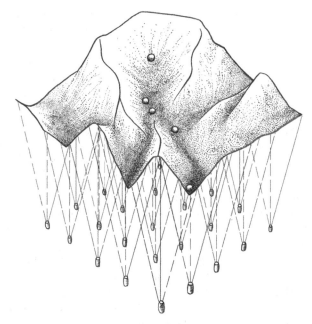

Figure 4.1 The Epigenetic Landscape (adapted from Waddington 1957). The development of a trait is likened to a ball rolling down an inclined surface. The surface has a series of valleys. As the trait develops it increasingly gets shunted into the valleys – canalised. Epigenetic (regulatory) relations between genes confer on the landscape its structure. Original artwork by Sylvia Nickerson.

perturbed, it would tend to move up the side of the valley, and thence be directed back to the proper trajectory. So, the topography of the epigenetic landscape secures the robust development of forms (see Figure 4.1). The topography of the landscape, in turn, is provided by the regulatory, epigenetic interactions amongst genes.[6] At times of significant perturbation, the developing chreode can be forced beyond the confines of the epigenetic valley, that is to say beyond the regime in which epigenetic interactions can buffer development against shock. In these circumstances strange phenotypes (at the time sometimes called 'phenocopies') can be produced.

Repeated selection for the production of these phenocopies, Waddington surmised, can alter the epigenetic landscape in such a way that the novel trait

---

[6] Waddington coined this idea before the discovery by Jacob and Monod (1961) of the mechanisms of gene regulation. On hearing of the Jacob and Monod account of gene regulation, Waddington immediately saw that it provided the mechanism for canalisation.

becomes increasingly canalised. In selecting the bithorax phenotype, Waddington supposes, we are also selecting the entire suite of epigenetic relations that contribute to its development. If the population exhibits variation in these epigenetic relations, then repeated selection and recombination can result in clusters of genes that are capable more reliably and robustly of producing what at one time was a highly unstable, exceptional phenotype.

Genetic assimilation is the process by which a phenotypic response to the environment becomes, through the process of selection, taken over by the genotype so that it becomes independent of the original environmental inducer. This idea had several predecessors, including those hypotheses of J.M. Baldwin, and is essentially the same as Schmalhausen's hypothesis of genetic stabilization. (Gilbert 2003a: 472)

So the Modern Synthesis response to this battery of putative counterexamples is to reaffirm the distinction between inheritance and development, and to reassert the commitment to hard inheritance. Insofar as there are genuinely *evolutionary* novelties occurring in these scenarios, they appear only after they become associated with new genes or genetic combinations that reliably produce them (Wallace 1986). Until such time, these novelties are mere 'ecological events' (in Sterelny's terminology). Their reproduction by the processes of development, or their inducement by the environment, may be sufficient to alter the process of selection in a way that makes the appearance of their corresponding genetic novelties more likely. But only when the latter event occurs do we have genuine evolution, because only then is the trait susceptible of 'hard inheritance'.

The Modern Synthesis response to the *prima facie* challenge to the Weismann doctrine raised by these phenomena is effective. But there is a further question how well motivated it is. There are well-documented instances of heritable, environmentally induced novelties that cannot be dismissed in this way. The insistence on Weismannism and hard inheritance may be pointless given the prevalence of these much more acute challenges.

It is known that environmental stress can induce alterations to the methylation pattern of DNA. Methylation occurs when a methyl group is added to a guanine or cytosine base on the DNA molecule. The methyl group regulates the transcription of the gene to which it is attached. A remarkable study of the persistent effects of the extended famine in Holland in the winter of 1944–1945, demonstrated that early exposure of human fetuses to famine conditions is associated with a decreased methylation of Insulin-like Growth Factor genes (Heijmans et al. 2008). These patterns are associated with diminished growth rates. These diminished growth rates in stressed females are eventually passed on to their offspring (Lumey 1992). Other studies demonstrate that methylation patterns induced in single-celled organisms, and in somatic cells of the metazoa and metaphyta, can be passed directly to daughter cells. Such epigenetic inheritance of acquired methylation patterns is especially well documented in plants.

While methylation patterns must be erased in the germline cells of metazoans, they nevertheless reappear in offspring through the mechanism of (primarily maternal) imprinting in which remethylation occurs after fertilisation (Ferguson-Smith 2011). They are not transmitted in the way that replicators are supposed to be. But it is clear that this process, nevertheless, can support the pattern of inheritance (Schmitz et al. 2013). These processes, then, provide 'additional sources of inheritance' (Echten and Borovitz 2013).

> We suggest that environmental induction of heritable modifications in DNA methylation provides a plausible molecular underpinning for the still contentious paradigm of inheritance of acquired traits originally put forward by Jean-Baptiste Lamarck more than 200 years ago. (Ou et al. 2012)

More and more, the importance of transmitted epigenetic markers is being realised (Guerrera-Bossagna et al. 2013). This too certainly looks like a violation of the independence of inheritance from developmentally induced changes of the sort supposed to be proscribed by the Weismann doctrine. And it cannot be explained away as gene replication or genetic assimilation.

Of course, the Modern Synthesis approach could simply accommodate epigenetic imprinting and inheritance, by extending its conception of a replicator: methylation patterns may be considered replicators (of a sort). The process of imprinting can even be brought under the ambit of the concept of replication.[7] This would require jettisoning the commitment to hard inheritance, but it would by no means be an illicit, or disreputably *post hoc*, manoeuvre on the part of defenders of the Modern Synthesis orthodoxy. Scientific theories ought to be permitted to make adjustments to their concepts in the light of empirical evidence, after all. But is not entirely clear how effective this concession might be. Abandoning the commitment to hard inheritance might signal a welcome degree of doctrinal pliancy on the part of the Modern Synthesis. But it may also be at once both inadequate, and a concession too far. It may turn out to be inadequate in the light of other further putative examples of nonreplicator mediated inheritance (Danchin et al. 2011). On the other hand, a wholly adequate conception of inheritance risks forcing a re-evaluation of the independence of inheritance from development, one of the cornerstones of the Modern Synthesis. I take these two points in turn.

---

[7] Sterelny (2001), for example, advocates an 'extended replicator theory' in which the category *replicator* includes cell structures, methylation patterns and some obligate symbionts.

4.3   Inheritance pluralism[8]

Once it is acknowledged that there are more ways to secure the intergenerational stability of form than through the transmission of replicators, the Modern Synthesis commitment to genes as the exclusive units of inheritance appears to lose some of its appeal. Control over the *pattern* of inheritance is not located in any one place. Rather it is 'causally spread' throughout the gene/organism/ environment system.[9] One increasingly popular way of giving voice to this idea of a plurality of inheritance mechanisms, is to enumerate discrete 'channels' through which inheritance is mediated (Lamm 2014).

Genetic inheritance: 'Genetic inheritance' refers to the process of replication and transmission of nucleotide sequences in DNA. It has come to be almost synonymous with 'inheritance'. It is certainly a powerful mechanism for the transmission of form. Indeed it seems ideally suited to the process. As nucleotide structure is stable, and copied with high fidelity, it is reliably passed from generation to generation. As DNA is modular each portion is buffered from the effects of changes elsewhere in the system, and, as it consists in a digital code, it is unlimited in its size and range of possible varieties (Szathmary 2000). These, recall, are the very properties that commended the 'order-from-order' answer to Schrödinger's question.

Epigenetic inheritance: The transmission and imprinting of epigenetic methylation 'markers' discussed above, looks to be a distinctly nongenetic mechanism for the passing on of crucial developmental resources. This appears to be a separate mechanism that operates independently of the replication of DNA.

Heritable epigenetic variation is decoupled from genetic variation *by definition*. Hence, there are selectable epigenetic variations that are independent of DNA variations, and evolutionary change on the epigenetic axis is inevitable. The only question is whether these variations are persistent and common enough to lead to interesting evolutionary effects. (Jablonka and Lamb 2002: 93)

In addition, cellular structures reoccur generation on generation without the mediation of replicated genes. 'Cortical inheritance', for example, has been demonstrated in paramecia. Errors and alterations in the cuticle of paramecia are passed on to daughter cells during cell division (Beisson 2011). Jablonka and Raz (2009) have documented more than 100 instances of epigenetic inheritance in 42 different species (Jablonka and Lamb 2010). It is no mere fringe phenomenon.[10]

---

[8] I take the term 'inheritance pluralism' and its alternate 'inheritance holism' from Mameli (2005), although my conception of holism departs significantly from Mameli's.

[9] The idea of 'causal spread' comes from Clark and Chalmers (1998).

[10] Gilbert and Epel (2009) have produced an impressive list of documented instances of epigenetic inheritance in animals. See their Appendix D.

Behavioural inheritance: Behavioural inheritance systems involve those in which offspring learn behaviours from parents or others. A classic example is to be found in the spread of the ability of great tits (*Parus major*) in Britain to extract the cream from milk bottles left on the doorsteps of houses. Further examples include dietary preferences that are transmitted to mammalian foetuses though the placenta.

Cultural inheritance: Languages, technological skills, social behaviours, cultural norms and traditions are all passed from one generation to the next through learning, the maintenance of written records and enculturation.

Environmental inheritance: Proper growth and development in mammals requires the correct gut flora. Failure to acquire the correct gut flora can lead to inability to digest food, immunodeficiency and morphological abnormalities. Many young acquire their gut flora from their mothers at birth as they pass through the birth canal. Parental nongenetic effects are very commonly transmitted. These may include traits that affect the development of secondary sex characters (Danchin et al. 2011). One intriguing example is found in birds. Many female birds confer on their offspring immunity to specific pathogens by placing specific antibodies in the yolk of their eggs. Hatchlings assimilate the antibodies and hatch with an immunity that has not been acquired in the normal way, but inherited from the mother. Agrawal, Laforsch and Tollrian (1999) demonstrated the transgenerational transmission of environmentally induced predator defences in both animals and plants.

Examples of 'extragenetic inheritance' are legion, and are widely known. They certainly appear to suggest that the pattern of resemblance and difference that constitutes our pretheoretical – or better 'pre-Modern Synthesis' – conception of inheritance is achieved by much more than the transmission of replicators. It may seem surprising, then, that the Modern Synthesis account of inheritance hasn't buckled under the weight of its empirical inadequacy. Yet, the conviction persists that inheritance is 'hard inheritance'. It isn't just obduracy that has kept hard inheritance in currency. As compelling as one might find the existence of multiple, nongenetic mechanisms of inheritance to be, they do not undermine the Modern Synthesis conception of inheritance as mediated exclusively through the transmission of genes (or replicators). The Modern Synthesis has the resources to withstand these challenges.

One possible response might draw on the (putative) distinction between evolutionary characters and nonevolutionary characters. This strategy might concede, for example, that these alternative modes of inheritance really are instances of genuine inheritance, but they are not the inheritance of *evolutionary* characters unless their passage from parent to offspring is mediated through replication.

An alternative strategy might be to concede, as Simpson (1953) appears to have done, and as Maynard Smith (1989) intimates, that alternative modes of inheritance are conceptually possible, and may in fact exist, but that they are rare and have little impact on evolution. This response may have been justified at one juncture in history, but it leaves Modern Synthesis evolutionary theory a hostage to empirical fortune. The possibility of significant amounts and kinds of nongenetic inheritance may have seemed remote in the 1950s, and perhaps even in the late 1980s, when Simpson and Maynard Smith were making their concessions. Now, however, the increasing prevalence of extragenetic inheritance is being documented at an impressive rate (Ferguson-Smith 2011; Gilbert and Epel 2009; Jablonka and Lamb 2010). It is abundantly obvious that this is not an aberration, or a curiosity, but a major influence on evolution to be reckoned with.

Yet another Modern Synthesis rejoinder might be to insist on the theoretically motivated distinction between the processes of development and those of inheritance. The pattern of resemblances and differences within and between lineages is not inheritance; it is the result of the combined activities of inheritance and development. The extragenetic resources that secure this pattern are parts of the process of development. Significantly, these include epigenetic and cell structures, features of the environment (like the presence of microbes to colonise the gut) and the organism's cultural setting. These processes – epigenetic, behavioural, environmental, etc. – are not distinct modes of 'inheritance' at all. They are simply different ways in which development contributes to the *pattern* of inheritance. Everyone concedes, of course, that development has a significant role to play in securing this pattern. But, so the response goes, the *pattern* of transgenerational resemblance and difference is *not* the *process* of inheritance.[11] To conflate the two is to forfeit the enormous theoretical dividend that accrues to the separation of development and inheritance made possible by the Modern Synthesis.

The point of the foregoing discussion has been to point out that while there are multiple processes that hold the *pattern* of inheritance in place – i.e., the transgenerational resemblances and differences – they do not necessitate wholesale changes in the Modern Synthesis conception of inheritance. Nor do they require the displacement of the gene as the privileged unit of inheritance. The reason is that the Modern Synthesis has its own proprietary definition of inheritance. It holds that any phenomenon that is not underwritten by the transmission of genes, or replicated materials, is just not inheritance. This commitment is justified and rightly earned by the success of the Modern Synthesis. A theory with this sort of pedigree has a right to stand its ground, especially against the charge that it fails to do justice to what is effectively a 'folk concept', or a pretheoretic intuition. The proprietary concepts of the Modern Synthesis have earned their credentials; they should not be given up lightly.

---

[11] This appears to be the strategy pursued by Dickens and Rahman (2012).

Whether these alternative channels for securing the intergenerational stability of form genuinely count as modes of inheritance or not, they do serve to accentuate how far the Modern Synthesis conception of inheritance as a discrete molecular process, independent of development, realised exclusively in the transmission of genes, has departed from its Darwinian precursor. There is much more to the persistent pattern of resemblance and difference within and between lineages, on which evolution depends, than is countenanced in the transmission of replicators. When inheritance as the theoretical process and inheritance as the observable pattern pull apart, it is just as reasonable to question the adequacy of the theoretical process, as it is to disregard the observable pattern. At this point, the defenders of the Modern Synthesis account of inheritance and its challengers reach an impasse. It isn't clear what might break it.

4.4   Inheritance holism[12]

An alternative approach is to question the coherence of the Modern Synthesis conception of inheritance on its own terms. The investiture of the gene as the unit of inheritance is predicated on a particular conception of the causal role of genes in securing transgenerational resemblance. The leading idea is that the effects of genes on phenotype can somehow be distinguished, or isolated, from the effects of extragenetic elements of the developmental system. Given that, genes can then be shown to be the principal difference makers for the evolutionarily significant resemblances and differences that hold within and between lineages. But, the challenge goes, given the complexities of gene function and development, this sort of causal decomposition is impracticable. The very idea that inheritance and development can be thought of as separate processes is based upon an erroneous conception of causal decomposition. I shall work up to this objection gradually.

4.4.1   The method of difference

The Modern Synthesis supposition that genetic and extragenetic causes of form can be differentiated is the legacy of a particular kind of causal reasoning, the *locus classicus* of which (in biology anyway) is Classical Genetics. In Classical Genetics, differences in phenotypes are correlated with – and hence attributed to – differences in classical genes. In Chapter 1, we encountered the role that Mill's Method of Difference plays in the 'analytic method'. It is equally crucial in securing the Modern Synthesis conception of genes as units of inheritance.

---

[12] This section draws heavily upon material published in Walsh (2014b).

Mill's Method works as follows. Suppose you want to know what contribution some putative cause, C, makes to some complex effect, E, you conduct an experiment, or an intervention, designed to demonstrate what would happen to E if C weren't there (Lipton 2004). Alternatively, you might manipulate the value of C – make it stronger or weaker – and observe what difference the intervention makes to E (Woodward 2003). In Mill's words

[t]he canon which is the regulating principle of the Method of Difference may be expressed as follows . . . :

    If an instance in which the phenomenon under investigation occurs, and an instance in which it does not occur, have every circumstance in common save one, that one occurring only in the former; the circumstance in which alone the two instances differ, is the effect, or the cause, or an indispensable part of the cause, of the phenomenon. (Mill 1843: 256)

The Method of Difference allows us to make inferences from the difference in effects in the absence of C to the causal contribution that C makes when it is there. Mill recognises that his 'Canon' of causal inference makes certain metaphysical assumptions, *viz.* that causes compose in an orderly way that permits their decomposition.

I shall give the name of the Composition of Causes to the principle which is exemplified in all cases in which the joint effect of several causes *is identical with the sum of their separate effects*. (Mill 1843: 243, emphasis added)[13]

That is to say, overall, the causes of the system are relatively insensitive to our interventions. In removing one cause, we leave the others more or less unaffected.

    While Mill acknowledges that some causes do not compose in this way, he expresses confidence that his principle of

the Composition of Causes, is the general one; the other [*i.e.* noncomposition] is always special and exceptional. There are no objects which do not, as to some of their phenomena, obey the principle of the Composition of Causes; none that have not some laws which are rigidly fulfilled in every combination into which the objects enter. (Mill 1843: 244)

The Modern Synthesis conception of the gene as an individual unit of inheritance was established through the applications of Mill's method. Waters (2007) offers an influential endorsement of the application of this mode of inference in developmental genetics.[14] Genes, Waters tells us, play 'an ontologically distinctive role' in phenotypes (polypeptide sequences at least) because they are '*specific actual difference makers*'. A specific actual difference maker is a

---

[13] We saw this supposition at work in the characterisation of mechanism.
[14] Northcott (2009) and Keller (2011b) offer critiques of Waters' argument that differ from mine.

cause that varies in a population, and whose variation correlates with a specific variation in effects. Because the variation in genes within a population corresponds with specific differences in phenotype, they are distinct and ontologically privileged causes.

Waters documents how Classical Genetics relied upon the notion of difference making in constructing its conception of the gene as a unit of phenotypic control and the unit of inheritance. Gene function was inferred through the application of what he calls the 'Difference Principle'

Difference Principle: differences in a gene cause uniform phenotypic differences in particular genetic and environmental contexts. (Waters 2007: 558)

The Difference Principle, he notes, was consciously deployed in the experiments of T.H. Morgan and his colleagues.

If now one gene is changed so that it produces some substance different from that which it produced before, the end-result may be affected, and if the change affects one organ predominatingly it may appear the one gene alone has produced this effect. In a strictly causal sense this is true, but the effect is produced only in conjunction with all the other genes. In other words, they are all still contributing, *as before*, to the end-result which is different in so far as one of them is different. (Morgan 1926: 305–306, emphasis added)[15]

This is simply an application of the method, complete with the explicit assumption of the Composition of Causes. Witness the claim that the other causes '. . . are still contributing, as before, to the end result'.

In some circumstances, the Method of Difference really is an effective means for identifying discrete, explanatorily privileged causes. Suppose we both shoot an arrow at a target and yours hits the bull's eye and mine misses. The cause of the difference between the outcomes is the velocity (direction and speed) at which our respective arrows left the bow. In that event, by dint of being the difference in causes that makes the difference in effect, the velocity of the arrow's trajectory enjoys a causal and explanatory privilege. It has causal privilege in that the velocity of the arrow is the causal factor that varies between the two events – your shot and mine – that also best accounts for the difference in outcome. There are, to be sure, other causes in play – the wind, barometric pressure, gravitational attraction of the Earth and other planets, etc. – but these are reasonably constant across the cases and more or less negligible, and so don't account for the difference in effects.

In this case, we can even say more. The velocity with which the arrow is shot not only best explains the *difference* between your shot and mine, it best

---

[15] This passage is quoted in Waters (2007), where it is deployed as part of an argument for the special theoretical privilege of genes.

explains your shot hitting the bull's eye. It is thus the privileged cause of your shot's outcome taken by itself.

If we are inclined to accept the velocity of the arrow as something like the satisfactorily complete causal story both of the *difference* in question (between your shot and mine), and the *effect* (say, your arrow hitting the bull's eye), taken in isolation, then we should be willing to accept that there is very little difference (if any) in this instance between the cause of a *difference* in effect and the ontologically privileged cause of the *effect* (Keller 2011b).

This seems to be just what Morgan is doing in the passage cited above. So, genes have a special privileged role in causing phenotypic *differences*, and consequently have a privileged causal/explanatory role in accounting for *phenotypes*, despite the fact, as Morgan readily acknowledges, there are many other causes, and that the full causal story is complicated. Morgan's form of reasoning suggests that the Method of Difference licenses the attribution of discrete, decomposable and privileged causal influence to discrete genes on discrete phenotypes.

In many instances the combined effects of co-occurring causes do not meet the conditions for decomposability. In cases where causes fail to compose in the right way, the best we can do is infer that the *difference* in an effect is due to the *difference* in a particular cause without being able to isolate the separable contribution of that cause.

Perhaps an example will help. Suppose we observe that for a given velocity, a rock of mass m doesn't break a window that it is thrown at, but a larger rock of mass $2m$ thrown at the same velocity does. Mass here is a specific actual difference maker, in Waters' sense. But it does not follow that the mass of the larger rock is an ontologically privileged cause of the window's breaking. The reason is that it is the rock's momentum that causes the window to break, and we cannot apportion the momentum into that part that is attributable to mass and that proportion attributable to velocity. Mass and velocity do not compose in the right way; momentum is their product. A difference in mass makes a difference to the capacity of the rock to break the window only insofar as it makes a difference to the rock's momentum. In this case, the method of difference only allows us to infer that the difference in a particular cause – mass – makes a *difference* in effect, without enabling us to infer the role of mass in causing the effect in question.

The significance of this for our purposes is that the experiments of Classical Genetics may allow us to correlate *differences* in genes with *differences* in phenotype, but they do not, thereby, allow us to demarcate the causal contributions of genes in normal circumstances, or to distinguish their causal contributions from those of the environment in the production of individual phenotypes (Keller 2011b). We can do this only if genes and extragenetic causes of phenotypes compose in the right way. But, generally they do not. Genes are components of complex self-regulating, adaptive networks, and these systems have a causal dynamics all their own. The architecture of these systems places some serious constraints on the kind and scale of causal decomposition that can be effected.

4.4.2   Causation in complex systems

Complex adaptive systems have the capacity to maintain a stable configuration in the face of perturbations precisely by altering the causal relations among their components. Such systems exhibit cyclic, top-down causal architectures. Each component in the system affects, and is affected by, the others. The overall effect of an intervention on the system's dynamics is *jointly* the result of *all the system's components*. Ultimately, the activities of each component of the system is affected by itself. This is, in effect, a radical failure of the Composition of Causes, and it impacts our ability to make the kinds of surgical interventions required by the Method of Difference (Mitchell 2008).

The various causes of the dynamics of complex adaptive systems are, in Wagner's (1999) terminology, 'nonseparable'. Causes are 'separable' in this sense, only if the effect of a change in one is independent of the effects of changes in others. That is equivalent to saying that we can intervene on one component while leaving others unaffected.

The failure of separability undermines the sort of causal decomposition on which the Modern Synthesis conception of gene function is predicated. It raises four major problems for causal inference. The first is that where causes are nonseparable, we cannot generally ascribe to any element of a complex adaptive system, any context insensitive causal role (Wagner 1996, 1999). Changing one component changes the activities of all others. So, causal activities of individual components typically do not generalise across contexts. The second is that, as in the case of velocity and mass, the several effects of the component parts are nonapportionable, in the sense that it is impossible to say how much of a given effect is attributable to one causal component and how much to another. The third consequence is that we cannot generally attribute *differences* in effect to specific *differences* in the causal contributions of discrete elements. This is because a difference in one causal component elicits a change in the activities of all or most components. Consequently, in such systems, one cannot assume, as Morgan does, that after an intervention on one component, the others '. . . are all still contributing *as before*, to the end result'. The fourth consequence – a corollary to the third – is that we cannot generally decompose differences in effect to those caused by influences external to the system from those internal to it.

This last consequence may need some elaboration. Changes in the dynamics of complex adaptive systems may be initiated endogenously through internal perturbations or exogenously through changes in their environment. In either case, the system mounts an adaptive response. Any such adaptive response is *both* a change in the internal behaviour of the system and a change in the way it interacts with its external environment. That response is *jointly* the consequence of the environment and the system's internal dynamics.

Feedback relations with the environment recalibrate the internal dynamics of complex systems to incoming signals. Doing so embeds the system in its contextual setting by effectively importing the environment into the system's very dynamical structure. (Juarrero, 2012: 2)

Complex adaptive systems raise a further difficulty for causal inferences. They make salient a distinction – less obviously important for linear systems – between what we might call 'principal causes' and 'initiating causes'. A principal cause is a cause to which a significant portion of an effect is attributable. An initiating cause, as the name suggests, is a cause that starts a causal process, which in turn, culminates in an effect. If two otherwise identical linear systems or processes diverge in their outcome, then (on the assumption that all other components are contributing 'as before') it is generally reasonable to ascribe principal causal responsibility for the difference in effect to the factor that *initiates* the different trajectories. Principal causes are just initiating causes. But the inference is less obviously appropriate for complex adaptive systems. It does not follow from the fact that a change in the dynamics of a complex adaptive system is *initiated* by, say, a change in external conditions, that the *principal cause* of the overall effect is that change in external conditions.

That's all a little complicated, but the take-home message is simple. Complex dynamical systems invalidate, or at least challenge, our usual procedures for inferring causes from effects. The reason is that compositionality – the 'canon' of our usual causal reasoning – breaks down so comprehensively in these systems. This has implications for the Modern Synthesis conception of the role of genes in inheritance and development.

4.4.3   Gene regulatory networks

The dynamics of gene regulatory networks are the object of an exciting research programme (Ciliberti et al. 2007a, 2007b; Davidson 2006; Davidson and Erwen 2006; Davidson and Levine 2006; Wagner 2011). One of the most striking and consistent results of this programme is that genes do not act as *units* of phenotypic control in the manner supposed by classical and early molecular genetics. The relation between individual genes and phenotypes is not a 1:1 mapping; it is inordinately complex (Wagner 2011).

Early knock-out experiments revealed a curious feature of gene action (Gibson 2002). Knock-out experiments employ a version of the method of difference: to find out what a gene does, see what happens when you disable it. Scientists noticed that, often enough, knocking out a gene failed to register any phenotypic effect. This, it became clear, is because genes ply their trade as parts of extremely complex, dynamic, self-organising networks (Ciliberti, Martin and Wagner 2007a; Davidson and Levine 2006; Wagner 2011). One of the principal characteristics of these networks is adaptive compensation. If one gene fails to work appropriately, the network alters its dynamics in a way that compensates, ensuring that the appropriate output is preserved (Meir et al. 2002; Wagner, 2012). This functional integration is achieved via a specific kind of control that the network as a whole exerts on the activities of each its component genes.

Top-down causation . . . is the ability of higher levels of reality to have a causal power over lower levels . . . Dynamic effects take place at some time, and the outcome would be different if the higher level context were different. Altering the high-level context alters lower level actions, which is what identifies the effect as top-down causation, where the high-level context variables are not describable in lower level terms, which is what identifies them as context variables. (Ellis 2012: 5)

It is in this context that we can see how inferences from *differences* in effect to specific principal causes break down. Gene networks famously damp out the effects of perturbations to (or manipulations of) individual genes (Greenspan 2001). For example, knocking out a gene may have no effect on the network's output. But that obviously doesn't license the inference that prior to the knock-out, the gene played no causal role at all (Mitchell 2008). Conversely, gene networks may amplify the effects of changes in gene action. A minor altera-tion to the action of a gene in a network may cause a huge difference in the network's output (Meir et al. 2002), but that does not mean that the principal cause of the new output is the intrinsic causal properties of the mutated gene.[16] When a novel phenotype is produced, the cause of the resulting phenotype may the adaptive response of the gene regulatory system *as a whole*. So, differences in phenotype cannot be attributed wholly to any particular causal component.

One consequence of this causal complexity is that a novel phenotype that is *initiated* by an internal perturbation – a mutation – cannot be said to be principally *caused by* that mutation. Likewise, a novel phenotype that is *initiated* by a change in environment cannot be said to be principally *caused by* that environmental change. This is because, in general, a phenotypic effect cannot be decomposed into the respective contributions of the system's internal dynamics and the environment. Every phenotypic novelty is thus equally 'genetic' and 'environmental', no matter how it is initiated. There is, then, no

---

[16] In fact, gene regulatory networks do not even licence the less demanding attribution of *differences* in effect to *differences* in specific causes.

difference in kind between phenotypic novelties *initiated* by mutations and those *initiated* by environmental perturbations. They are all jointly caused by the entire complex adaptive system

The upshot is that where genes operate as parts of complex regulatory networks, reliance on the Method of Difference to infer – or apportion – their several causal contributions to inheritance can be wildly misleading.

> Identifying genes by *differences* in phenotype correlated with those in genotype is therefore hazardous. Many, probably most, genetic modifications are buffered. Organisms are robust. They have to be to have succeeded in the evolutionary process. (Noble 2008: 3007, emphasis in original)

The nonseparability of causes doesn't entail that we can say nothing about the causal roles of genes. Though the causal roles of genes in organisms may be undissociable, and undifferentiable because of their enormous context sensitivity, we may nevertheless be able to pick out regular, projectable causal relations in more restricted contexts (Strevens 2005).[17] In extremely dense systems of nonlinear circular causation and feedback, the best strategy is to concentrate on highly localised causal relations. Thus, for instance, the role of a gene within a gene regulatory network may be reasonably invariant. We may, for example, be able to discern that a transcription factor fairly consistently up regulates, or down regulates, the transcription of a particular structural gene. Alternatively, we may discover that a discrete region of DNA regularly influences the construction of precursor mRNA. In this way localised causal relations can be built up piecemeal.

I take it that this process of highly local, causal analysis is the way that gene wiring diagrams are generated (in the manner of Davidson 2006). The important point is that this process is substantially different from that of tracing the large-scale effect of a single gene on a phenotype. In strongly nonlinearisable complex systems, local causal relations are not generally susceptible to long-range extrapolation. So even though we can say what role a gene plays in a gene network, that in no way entails that we can ascribe to it a definitive role in the production of a phenotype.

4.4.4   Genes in inheritance

The causal structure of gene-regulatory systems has far-reaching implications, some of which will be explored in the coming chapters. For present purposes, it allows us to draw some tentative conclusions about inheritance. It suggests that inheritance is holistic in a way that completely undermines the traditional Modern Synthesis view of individual genes as discrete units of inheritance.

---

[17] I thank Cory Lewis for helpful discussion here.

In the process it severely compromises the Modern Synthesis insistence on the discreteness of inheritance and development.

The traditional picture is predicated upon four metaphysical commitments concerning the relation between genes and extragenetic factors. These are: (i) the causal contribution of genes or germline entities to phenotypes are discrete and relatively context insensitive; (ii) there is a difference in kind between those novelties whose principal causes are genetic (germline) changes, and those whose principal causes are extragenetic; or *at least* (iii) novelties, or differences in phenotype, that are *initiated* by genetic (germline) differences differ in their inheritability from novelties (or differences) *initiated* extragenetically; and (iv) the causal asymmetry between genotype space and phenotype space induces a conceptual asymmetry, such that only changes in genotype space are genuinely evolutionary.

Inheritance holism contends (*contra* [i]) that the respective roles of genes and extragenetic factors in inheritance cannot be disentangled, because they cannot be causally decomposed. Furthermore, the contribution that an individual gene makes to a single phenotype is not constant; it varies wildly according to the context it is in. Consequently, (*contra* [ii]), every trait, including every novel trait, is a result of the complex, adaptive causal interaction between genomes, organisms and environments. There is, then, no *evolutionary* difference in kind between those traits *initiated* by a change in genes (mutation) and those initiated by extragenetic factors. Every phenotypic novelty is the result of the assimilation, regulation and accommodation of *all* the causal influences impinging on the reactive, adaptive genome. Because there is no difference in kind between genetically and environmentally induced novelties, (*contra* [iii]) there is no difference in kind in their capacity to become intergenerationally stable, that is to say 'inherited'.

Finally inheritance holism claims (*contra* [iv]) that the conceptual asymmetry between genotype space and phenotype space fails to hold. It fails precisely because of the adaptive, reactive nature of gene regulatory networks and genomes. The Central Dogma of Molecular Biology says that processes downstream of sequestration (or of transcription) do not affect the germline (or the *structure* of genes). That, in turn, is thought to establish the primacy of genotype space over phenotype space. Whether or not the dogma is true, it is irrelevant. The Central Dogma cannot do the work that the Modern Synthesis requires it to do. The *activities* of germline entities (genes) – their regulatory relations, the timing and rates of action – are heavily dependent on the adaptive interaction of genomes with their downstream contexts. What genes *do* in the production of phenotypes is highly dependent upon developmental, behavioural, social and environmental conditions (Moczek 2012), even if their structures are not.

Thus, according to inheritance holism, genes are not the units of inheritance. The pattern of inheritance is held in place by the self-regulating, adaptive activities of organisms embedded in their environments. Genes have an important role to play in the reliable production of phenotypes, but it is not a role that can be differentiated, and detached, from the activities of any other components of the system.[18] Nor is it a role that earns genes any privileged theoretical status. Explaining the pattern of inheritance requires us to take into consideration this complex of commingled causes. There are two crucial implications of holism for the Modern Synthesis conception of inheritance. The first is that the process of development is intimately involved in the process of inheritance. There is no distinction between them, as the Modern Synthesis insists. Danchin and Pocheville echo the view that extragenetic modes of inheritance render the Modern Synthesis separation of development, inheritance and evolution unworkable.

[T]he emergence of *nongenetic inheritance* is providing a unique way of bridging physiology and evolution a link that remained quasi impossible as long as we persisted in reducing *inheritance* to its sole genetic dimension .... (Danchin and Pocheville 2014: 2308)

The second is that the Modern Synthesis notion of inheritance – transmission of replicated entities – is completely inadequate to explain the pattern of inheritance that evolution by natural selection requires (Mesoudi, Blanchet and Charmentier 2013).

The Modern Synthesis conviction that inheritance is mediated exclusively by genes is predicated on what we saw Nancy Cartwright (1999) describe as the 'analytic method' in which it is presumed that the parts of a complex system act, in all contexts, 'according to their natures'. This is the simple, powerful method that undergirded the mechanism of the Scientific Revolution. But this method, it turns out, is not universally applicable. In fact, it is rather parochial. It breaks down where the elements of a complex system do not have context insensitive causal capacities ('natures'). These are precisely the kinds of systems in which genes operate. In such causally dense, cyclical causal systems as organisms, characterised by top-down regulation, by highly context sensitive activities of parts, one cannot simply isolate the contribution of one component on the overall effect, and distinguish it from that of other components.

Yet, the supposition that one can is the motivation behind the Modern Synthesis fractionation of the *pattern* of inheritance – the transgenerational resemblances and similarities – into two discrete causal components, one

---

[18] Patrick Bateson seems to have been one of the earliest and most persistent advocates of the view that genetic and extragenetic influences on form are so complex that they cannot be decomposed or their respective effects apportioned. See, for example, Bateson (1976).

originating from the transmission of replicators, the other from development. This separation of inheritance from development is a fundamental tenet of the Modern Synthesis, but it is unsustainable. The nonseparability of inheritance and development is due not to the fact that the pattern of inheritance is causally 'spread', as pluralists would have it, but because it is causally nondecomposable, as inheritance holism insists. The principal casualty of this nonseparability is the concept of the single gene as a discrete unit of inheritance.

Conclusion

The notion of the gene or replicator as a unit of inheritance is a powerful and compelling one. It is predicated upon a particular conception of causal inference, according to which specific differences in phenotypic effect can be traced and attributed to specific differences in individual genes. Privileging genes as units of inheritance presupposes that the effects of genes on phenotypes can somehow be differentiated from the effects of the rest of the developmental system. Unfortunately, the relation between individual genes and phenotypes just doesn't permit that.

This much has been revealed by recent advances in our understanding of gene function and genome architecture. But, conceptual change has lagged behind empirical advances. It is time to ask how this enhanced understanding of genes reflects upon the conceptual underpinnings of the Modern Synthesis. The concept of biological inheritance is the ideal place to start this re-evaluation. The concept of inheritance does not enjoy a life of its own. It was forged with the theory of evolution, and has adapted to that theory's needs. As evolutionary thinking changes, so too do the demands it places on the conception of inheritance. The Modern Synthesis interpretation of inheritance may even have been instrumental in delivering biology to the point where it has the empirical resources to reassess its own foundational commitments. It is far from clear that the proprietary Modern Synthesis conception of inheritance will continue adequately to serve the needs of evolutionary thinking. It seems likely that the gene as the unit of inheritance, and the independence of inheritance from development, may be among the first Modern Synthesis precepts to fall. As we shall see in subsequent chapters, as these pillars give way, they threaten to take the other stays of the Modern Synthesis with them.

5   Units of phenotypic control
Parity or privilege?

The understanding of gene structure and function that arose in the twentieth century grew up alongside the development of both the digital computer and communication theory (Kay 2000). These fields not only offered elegant metaphors for the growth of organismal form, they also provided a guiding research strategy. The structure of DNA was to be studied as a language or a code; the objective was to decipher its message. The genome became a program, or blueprint, that specifies the features of the organism to be produced. Under the auspices of this metaphor, development becomes computation; the process in which the instructions encoded in the genes is implemented. The metaphor certainly has some influential exponents.

> What are transmitted from generation to generation are the 'instructions' specifying the molecular structures: the architectural plans of the future organism. . . . The organism thus becomes the realization of a programme described by its heredity. (Jacob 1973: 1–2)

It is a powerful and vivid image that at once makes the recondite details of development and inheritance somewhat more transparent and underscores the privileged status of genes as units that exert a specific and powerful control over the phenotype. It further gestures toward a computational/representational conception of living things, the organism as simply the output of a process that reads its instructions from a program. But, as G.C. Williams (1966) claims, what is important to evolution is the persisting program. Jacob continues (from above): 'An organism is merely a transition, a stage between what was and what will be' (1973: 2). Such ephemera are not the proper object of evolutionary study. Evolutionary biology, properly construed, addresses itself to the origin and persistence of the program or code, rather than its implementation.

> Developmental biology can be seen as the study of how information in the genome is translated into adult structure, and evolutionary biology of how the information came to be there in the first place. (Maynard Smith 2000: 177)

Here again, we encounter the influence of the order-from-order approach to understanding organisms. Erwin Schrödinger, as we saw, predicted that the molecular basis of life would be revealed to take the form of a 'codescript'.

'The chromosome structures . . . are the law-code and executive power – or to use another simile, they are the architect's plan and builder's craft in one' (Schrödinger 1944: 20). His 'aperiodic' crystal would have to have the features of a code that would permit it to record and retain a practically unlimited amount of information. That in turn would require it to have certain structural features. It must be easily copied, easily read, stable and open-ended. Indeed this conception of a molecular code was one of the central ideas guiding Watson and Crick's search for the structure of DNA. The generic features of a code, that are so applicable to DNA are: (i) Arbitrariness: the relation between a code, c, and the feature of the world that it encodes, w, should be arbitrary in the sense that it is not a matter of nomological necessity that c co-occurs with w. (ii) Stability: the code must be capable of maintaining its structure over long periods of time, and it must be resistant to perturbing influences. (iii) Modularity: it must be the case that changes or additions to the code in one place are isolated from the structure of the code in other places. In this way changes and novelties can accumulate the feature of (iv) open-endedness: the code must not be limited in its length. It must be capable of making up a string of arbitrary length. (v) Productivity: it must be possible to code any number of new messages, derived from the existing sequence (Maynard Smith 2000).

Of course the structure of DNA meets these specifications admirably (Szathmary 2000). It is hardly surprising, then, that the gene-code metaphor should have such currency. In fact, it is plausibly not a metaphor at all.

The analogy between the genetic code and human-designed codes such as Morse code or the ASCII code is too close to require justification. (Maynard Smith 2000: 183)

It seems reasonable to suppose that DNA does indeed, in a very literal sense, encode information (Godfrey-Smith 2000).[1] In transcription, RNA polymerase reads the nucleotide sequence and builds a corresponding (precursor) messenger RNA, which leads eventually to the construction of a protein, comprising a folded sequence of amino acids. But the fact that DNA is a code no more secures its control over the phenotype than the fact that the alphabet is a code secures for it a privileged control over poetry. That DNA is a code does nothing to support the genetic program metaphor.

Nevertheless, the computer program metaphor is a compelling one.[2] It nicely elicits the idea of genes occupying a governing role in the production of an organism, analogous to that occupied by a program in the production of some computational output. The principal challenge has been to give the metaphor some substance. Fleshing out the metaphor involves two things: (i) specifying what it is about genes and programs that makes the former relatively like the

---

[1] Griffiths and Stotz (2013) call this 'Crick' information.
[2] Not everyone agrees. Nicholson (2014) claims that it is 'incoherent'.

latter, and (ii) saying why being relevantly like programs should confer on genes or genomes the distinctive theoretical privilege accorded to them in Modern Synthesis theory.

5.1   Privilege or parity?

Given the importance to the Modern Synthesis of genes as privileged units of phenotypic control, it is hardly surprising that its guiding metaphor has been subject to particularly heavy scrutiny. There exists a range of proposals for securing privileged place of genes in the generation of form. They encompass claims that genes have a unique causal role in the production of phenotypes, or that genes carry information, or that genes alone represent phenotypes. Each of these, in turn, has been strongly rebutted. The alternative to privileging genes is to accord them parity with other influences on development and inheritance.[3]

5.1.1   Causation

One prominent line of thought is that genes earn their special status on account of playing a distinctive causal role in the production of phenotypes (Waters 2007; Weber forthcoming). Opponents are quick to point out that control of inheritance and development is distributed throughout the organism/environment system (Keller 2011b; Oyama 1985; Oyama, Griffiths and Gray 2001; Stotz 2006a, 2006b). Even the function of individual genes is strongly regulated by their epigenetic settings.[4] There are, in addition, epigenetic, organismal, environmental and cultural influences on form, each of which has some claim to special importance; none of which lays an exclusive claim to explanatory privilege.

One way to motivate this causal parity is to point to a venerable, if underappreciated tradition in evolutionary biology that sees the problems of the generation and perpetuation of form as problems to be solved not so much by studying the activities of genes exclusively, but by addressing the physics of biological materials.[5]

Cell and tissue, shell and bone, leaf and flower, are so many portions of matter, and it is in obedience to the laws of physics that their particles have been moved, moulded and conformed ... their problems of growth are essentially physical problems, and the morphologist is *ipso fact* a student of physical science. (Thompson 1961: 7–8)

---

[3] Stegman (2012) gives a helpful taxonomy of parity theses.

[4] Stotz (2006a) offers a particularly nice summary of the epigenetic factors that control gene expression.

[5] The tradition has been strongly revived by Stuart Newman and colleagues. Vivid examples are to be found in Newman (2011, 2014, 2015), Newman and Bhat (2008) and Forgacs and Newman (2005).

The dynamics of development, the patterns of inheritance, the generation of phenotypic novelties, are the consequence of 'generic' physical processes that apply to organisms in virtue of their construction out of cells and tissues.

If physics plays a causal role, along with genes, in the development of organismal form, there must be aspects of development and its outcomes ... that are 'generic' and are explicable in terms of processes that living systems have in common with nonliving ones.... Animal embryos and regenerating tissues and organs consist of viscoelastic materials that are simultaneously chemically or mechanically excitable: they are capable of storing energy and thus responding to disturbances in an active rather than a passive fashion. (Newman 2014: 2404–2405)

The generation of form is both 'generic' and genetic (Newman 2015). The role of genes in the control of form can often only be understood in the context of the extragenetic physical systems in which they participate.

Specific gene products in the developing embryo help to mobilize different physical effects – surface tension, viscosity, elasticity, phase separation, solidification – and the evolution of developmental regulatory genes cannot be understood apart from the physical effects they directly or indirectly mobilize. (Newman 2011)

To attribute the control of form genes alone is to underestimate the complexity of life.

Examples of the ways in which the influences on form are causally spread are legion. Dynamic Patterning Modules (Hernàndez-Hernàndez et al. 2012; Newman and Bhat 2008, 2009), for example, are collections of genes that operate in combination with morphogenetic structures to control, and coordinate the development of cell and tissue structures, like tissue layering, blastula formation, segmentation, etc. Here the viscoelastic properties of cells and tissues are just as important in the production of biological structures as are the activities of genes.

Maternal effects, like the determination of the antero-posterior axis and the segmentation of *Drosophila*, are implemented by the interaction of the developing embryo with protein gradients in maternally contributed egg cytoplasm. In mammals proper digestive, endocrine, immune, and even cognitive development is dependent upon young assimilating the appropriate bacterial gut flora (Gilbert and Epel 2009). This flora is transmitted to the fetus as it passes through the birth canal. Antibodies are passed from mother to fetus and mother to newborn (Zinkernagel 2001). In addition there are multiple examples of the determination of form by environmental factors. Water nymphs of the genus *Daphnia* have two forms: one 'normal' form, the other with an elaborately elongated and horned hood. The hooded form is induced by the presence of predators in the environment of the developing *Daphnia* (Moczek et al. 2011).[6]

---

[6] The induction of this morph by the environment is transmitted across generations.

The general point is that the control of phenotypic form is distributed throughout the entire gene/organism/environment system, in a way that makes attributing causal priority to genes implausible and misleading. Genes should thus be accorded causal parity with other influences on form.

5.1.2   Information and instruction

It is not entirely clear that considerations of causal parity are decisive in this issue. Causal parity may be consistent with the sort of *explanatory* privilege that has been extended to genes by the computational/representational conception of the organism. The special role for genes, on that view, derives from the putative fact that genes *uniquely* encode *information*, or an *instruction* for building an organism (Stegmann 2005). Causal parity may not imply informational or instructional parity. For example, we may not be able to decompose the output of a computer program into those parts uniquely caused by the software and those uniquely caused by the hardware. Nevertheless, the software/hardware distinction is a significant and explanatory one. A computer program is explanatorily privileged (at least for some explanations of a computer's output), even if it is not causally privileged. So, causal parity notwithstanding, genes may still have explanatory or theoretical privilege on account of embodying information, or a program, or a set of instructions for building an organism, just as the computational/representational conception supposes.

Information and meaning   Information is a relational concept: a signal is said to carry information about a source. The initial formal theory of information was developed by Claude Shannon and Warren Weaver in 1948. Their account envisages a sender transmitting a signal to a receiver through a channel. Intuitively, if there is no correlation between a putative source and its signal, then the receiver can infer nothing about the source from the signal. The signal, such as it is, conveys only noise. But, if the features of the signal correlate with the state of the source, then one can infer the features of the source from the signal. The signal thus carries information about the source.

Clearly, genes carry information about phenotypes, in this formal sense. But the Shannon-Weaver approach to information is inadequate to establish a privileged status for genes, for a number of well-rehearsed reasons. For one, it fails to demonstrate that genes carry information uniquely. In fact, it entails that they don't. The trouble is that the relation of covariation that is used as an index of the amount of information is not asymmetrical. So, just as certain of my genes carry information about the colour of my eyes, the colour of my eyes carries information about certain of my genes.

John Maynard Smith recognises the inapplicability of this statistical notion of information to the computational/representational conception of the organism. Instead, he proposes that genes carry information about phenotype in a much more robust, metaphysically committing sense, analogous to the way in which words bear meanings, or thoughts are about features of the world. Genes have semantic content; they *represent* phenotypes:

[T]he concept of information is used in biology only for causes that have the property of intentionality.... A DNA molecule has a particular sequence because it specifies a particular protein, but a cloud is not black because it predicts rain. This element of intentionality comes from natural selection. (Maynard Smith 2000: 189–190)

Maynard Smith's argument is that genes have intentional (i.e., semantic) content because each particular gene has been promoted in a population by selection because of its capacity to produce a particular trait (Millikan 1984, 1989a). This teleosemantic approach to gene content has been given a further, sophisticated extension by Shea (2007, 2011).

Genes may well represent traits, but they are not alone in doing so. The alternative inheritance systems discussed in the preceding chapter – epigenetic inheritance, cytological and cultural inheritance – have presumably been promoted by natural selection for their capacity to contribute to the successful production of phenotypes across generations. So they too might be said to carry 'semantic information' about organisms' traits (Griffiths 2001; Sterelny 2000b). Whatever claim genes might make to exerting a unique control over the phenotype, it does not reside in their 'teleosemantic' content.

The failure of these attempts to articulate a genuine notion of 'genetic information' or genetic content that might ground the claims of a privileged theoretical role for genes has caused some philosophers, at least, to despair of ever finding one.

[T]here is no clear, technical notion of 'information' in molecular biology. It is little more than a metaphor that masquerades as a theoretical concept and ... leads to a misleading picture of possible explanations in molecular biology. (Sarkar 1996: 187)

Paul Griffiths extends the pessimistic assessment; he calls the information paradigm 'A metaphor in search of a theory' (Griffiths 2001). I think, perhaps, there is such a theory.

5.2   Natural imperatives

I opened this chapter with the rather nebulous claim that the 'genetic program' metaphor lends gene priority a certain intuitive appeal. It is reasonable to think of these various accounts of the presumptive causal, informational, or semantic properties of genes as attempts to give that metaphor some concrete meaning. But they are undecisive for three reasons. The first (and least significant) is parity.[7] In each case it is far from clear that genes uniquely possess the causal, informational or semantic properties attributed to them. The second is that very little is said about how being privileged in respect of, say, carrying information, or being causally special, or having representational content, should make genes relatively like computer programs. The third is that none of these attempts succeeds in specifying the sense in which being relevantly like a computer program might underwrite the special theoretical privilege that genes enjoy. That is to say, none of them captures what is intuitively appealing about the metaphor. What is it about computer programs that, *should* genes be relevantly like them, they would thereby lay claim to some particularly privileged place in the explanation of organismal form? I think we can answer this question by appeal to an informal conception of information. That is, we can say in informational terms what it might be for genes, or genomes, to be programs, that would confer on them their presumptive theoretical status. Then we can ask whether they are.

Take a line of computer code, for example, a 'PRINT' command in BASIC. This command plays a causal role in producing the output, it carries information about the output, it even represents the output. But it is not its causal, informational or semantic properties *per se* that confer on this line of code its special explanatory status. Commands are unique and privileged in explaining computational outputs because they are *commands*. A command carries imperative force: 'Print this'. In doing so, it asserts a particular control over the output. Nothing else in the software-hardware-environment system does that. So what confers on a line of code its explanatorily privilege is its illocutionary force.

Plausibly, then, what makes genes (or genomes) relevantly like programs is that they, and they alone, issue commands to developmental systems: 'Build phenotype *P*.' That would certainly make sense of the 'gene program' rhetoric, and it would explain their presumptive theoretical privilege. But if so, the viability of the program metaphor incurs a significant risk. A line of code (or an entire program) gets its imperative force from the intentions of the programmer. Organisms and genomes don't have intentional programmers. So, for the analogy to go through, there would have to be some nonintentional account of imperative illocutionary force. Happily, there is. It arises out of certain game-theoretic approaches to natural meaning.

In his seminal work on natural meaning, David Lewis (1969) wonders how signs come to stand for things in the world. The relation between words and the world is arbitrary after all (just like the 'gene code' is supposed to be). So how does this connection come about: how, for example, does 'cat' come to signify

---

[7] This is the least significant because genes could conceivably have parity with extragentic elements, in any or all of these senses, and still be theoretically privileged.

cat? Lewis' answer is that users of a language have to agree upon some convention: this word, 'cat', should be associated with this object, a cat. He likens communication to a cooperative game in which (let's say) there are two players – a sender and receiver – and they must co-ordinate to produce the optimal outcome for each.

Lewis illustrates the idea with a helpful (albeit slightly apocryphal) historical example. In 1775 the British troops were descending on Boston to quell the nascent uprising. Paul Revere, famously, had to get word to the garrison about the approach of the British. He could not see whether they were approaching by land or by sea. The sexton in the Old North Church tower could see, but could not ride to the garrison. So they hatched a plan. They agreed that the sexton would raise one lantern if the British approached by land and two if by sea. If Revere saw one lantern he would instruct the garrison to defend the land. If he saw two, he would instruct the garrison to defend the shore. The best outcome for both Revere and the sexton would be for Revere to instruct the garrison to defend the land were the British to approach by land.[8] In this instance, Lewis suggests, the 'one-lamp' signal takes on meaning. One lantern literally 'means' the British are approaching by land.

Such conventions can arise through the explicit agreement of the participants, of course, but as Lewis demonstrates they might come about *ab initio*, without the prior use of language. When the sender issues signal s_1 in situation w_1 and the receiver issues action a_1 on receipt of s_1, and given that in situation w_1, a_1 produces the best outcome for both sender and receiver, then, Lewis contends, this strategy will be game-theoretically stable: s_1 will come to mean w_1, whether or not there has been explicit prior agreement.

Lewis' game theoretic approach to conventional meaning doesn't deliver all that we need. It doesn't establish how a natural sign could come to issue a command. It does not distinguish a message with declarative force, from one with imperative force. For example, the one-lamp signal could be thought of as having declarative content: 'the British are coming by land'. But it could also be thought of as conveying an imperative: 'inform the garrison to defend the land'. Indeed, this signal system arguably only succeeds if it both declares a state of affairs in the world and issues a command.

Kevin Zollman (2011) has taken up the challenge of specifying the conditions under which a signal can be said to have strictly imperative, rather than declarative, illocutionary force. He argues that what is needed is 'informational asymmetry'. In a two-player game, such as the one that Lewis outlines, in which a signal is issued in response to some state of the world, w, and induces an action, a, the signal has declarative force 'w obtains' – and does *not* issue

---

[8] And to defend the shore if the British approach by sea, but for now we shall just concentrate on the conventional meaning of the 'one-lamp' signal.

an instruction – 'perform *a*' – if we can infer more from the signal about the state of affairs, *w*, than we can about the appropriate action, *a*. Conversely, the signal carries imperative force – issues a command – if it allows us to glean more about the appropriate action, *a*, than about the sate of affairs in the world, *w*. The key to distinguishing declarative from imperative force, then, is this informational asymmetry. Specifically:

Declarative: Pr(state|signal) > Pr(action|signal)
Imperative: Pr(action|signal) > Pr(state|signal)[9]

The following example might help to illustrate the idea. Two players, *S* (the Setter) and *D* (the Director) are to cooperate in arranging four coloured blocks: two red and two black. The objective of the game is to place all four blocks in a line in alternating colours – either *Red, Black, Red, Black*, or *Black, Red, Black, Red*. The blocks start off in a randomly chosen sequence, and they must be adjusted, or left, as the case demands. The Setter, *S*, can touch the blocks, but because he is wearing coloured lenses cannot distinguish between the black and red blocks. The director, *D*, can distinguish between the blocks, but, as she is too far away, she cannot arrange them. So they must work together.

Let's suppose that the starting arrangement is *B,R,R,B*. There is a range of things that *D* can say to *S* in order to produce the desired result. *D* could say (for example): '*Switch blocks one and two*', or '*Switch blocks three and four*'. Clearly these are commands, and not declarative statements. The informational asymmetry criterion captures this. If *D* *says* '*Switch blocks 1 and 2*' then *S* would know (and we would know) what the appropriate action would be in response. But *S* would not know from *D*'s statement alone (nor would we), what the initial arrangement of the blocks is. The command '*Switch blocks 1 and 2*' would be appropriate in either of the starting arrangement *B,R,R,B* or *R,B,B,R*. While this is an effective command – an imperative – it tells us less about the state of affairs to which the command is an appropriate response. Therefore, by the information criterion, it has imperative, but not declarative content.

But *D* could do something else. She could report on the starting order of the blocks. By saying '*Black, Red, Red, Black*' she ensures that *S* has sufficient information to know what to do. This message clearly has declarative content. But it strictly underdetermines the course of action that *S* should take. He might exchange blocks 1 and 2, or alternatively rearrange blocks 3 and 4. There is informational asymmetry here. The locution tells us more about the state of the world than it does about the elicited action. So, according to Zollman's

---

[9] These probabilities are subjective probabilities. The conditional probabilities can be thought of as how much information one can infer about the state of affairs or the appropriate action, given the signal.

informational asymmetry criterion, D's statement carries declarative, but not imperative force.

Note that there is a sense in which S and D have causal parity in arranging the blocks. Typically, neither contributes more to a particular outcome than the other. And their contributions are nondecomposable. In general, we cannot say *how much* of a particular outcome is due to one of the players and how much to the other. Nevertheless, despite the causal parity, we can say that D plays a special theoretically privileged role on account of the fact that she sends a specific command to S. Where imperative content is involved, theoretical privilege does not require causal privilege. Similarly, D's statement and S's action carry information. S's statement has semantic content, no matter what influence it has on the outcome. But whether D's statement exerts specific control over the implementation of S's action depends upon its illocutionary force.

The information criterion allows us to distinguish conventional signs that carry imperative force from those that carry declarative force. Here the concept of information is instrumental in accounting for the allure of the genetic program metaphor. Genes, according to the metaphor, are relevantly like programs in issuing commands for the construction of phenotypes. The representational/computational conception of the organism may be viable after all. At least now, it appears, we have a criterion against which to test the intuition.

5.3   Genes or developmental systems: who gives the orders?

Think of the setup in the following way, genes and extragenetic developmental systems are players in a co-ordination game. The optimal payoff to each player is the reproduction of a viable phenotype, given the conditions under which they jointly operate. One of the players, Player 1, can be said to be sending signals with imperative force to the other, Player 2, if Player 1's response to the conditions in the world induces Player 2 to effect some actions – such as produce a specific phenotype – and Players 1's response carries more information about the appropriate action than about the conditions to which it is a response.

This particular informational criterion appears to make sense of at least some of our talk of 'genes for' particular traits. Consider the genes for sex-linked phenotypes haemophilia or colour blindness. Males inheriting a copy of such a gene robustly develop the phenotype in question: colour-blindness or haemophilia. Let the gene be Player 1, and the developmental system downstream of the gene be Player 2. The gene can be said to send a signal, s, with imperative force 'make phenotype a' in circumstances, c, to the developmental system, just if we can infer more about the signal (the gene) from the phenotype than we can about the circumstances under which the phenotype develops.

Fairly obviously, and fairly trivially, in such cases the informational asymmetry obtains. The haemophilia and colour-blindness genes encode a program for building their respective phenotypes.

We tend to speak of a 'gene for' some particular phenotype when the correlation between the genotype and phenotype is robustly insensitive to context. Under these circumstances, I submit, it genuinely does makes sense to say that the gene embodies an instruction for constructing the particular phenotype, on the grounds that genes carry a certain kind of information. It applies quite generally to a class of phenomena that John Dupré calls the 'error in the genome' concept of the gene (Dupré 2012: 267). Arguably, any trait that robustly conforms to Mendel's pattern also meets the informational criterion for encoding a command. Moss (2003) reserves the term 'Gene-P' for genes that instantiate this relation. It isn't much of a stretch to suppose that the experimental protocol of Classical Genetics was designed (inadvertently) to capture instances of this schema.

But the fact that there are some genes for some phenotypes is not sufficient to establish the theoretical priority that genes have enjoyed in evolutionary theory. That privileged status requires, at least, that the informational asymmetry between genes and developmental systems holds in general. Here, it seems that recent empirical advances in the study of development, particularly in developmental genetics are germane, and stand to threaten the presumptive primacy of genes.

The most remarkable feature of organismal development that has come to light in the past twenty years is its supple, self-organising, adaptiveness (Kitano 2004). The processes of development respond to perturbations in ways that preserve and maintain the organism's viability across a wide array of conditions. Each gene, or each gene network, has a 'phenotypic repertoire', a wide range of outputs that it can produce across a range of circumstances. Which element of a phenotypic repertoire that a gene or gene network produces on an occasion is determined largely by the regulatory influence of the developmental system. This has a number of implications for evolution. I shall canvass a couple of these before returning to the 'who's giving the orders?' question. The degree to which organisms are robust, flexible, self-regulating wholes, will turn out to inform the issue of whether we should accept the idea that genes are issuing imperatives.

5.3.1   Orchestration

The evolution of complex adaptations requires the co-ordination of all of an organism's developmental systems. An evolutionary change in, for example, the length of a forelimb requires concomitant changes in bone deposition, muscularisation, innervation, circulation. Each of these systems must respond in ways that adjusts to the changes in the others. Organisms adapt during their development in order to accommodate phenotypic novelties. The plasticity of development secures this ability. Mary Jane West Eberhard calls this 'phenotypic accommodation'.

Phenotypic accommodation due to phenotypic plasticity is the immediate adaptive adjustment of phenotype to the production of a novel trait or trait combination ... Phenotypic accommodation reduces the amount of functional disruption occasioned by developmental novelty. (West-Eberhard 2003: 147)

Phenotypic accommodation of this sort helps to co-ordinate an organism's various developmental systems in response to a demand imposed by novel phenotypes. Accommodation appears to be a precondition of the evolution of highly complex adaptations. If each system required its own independent genetic mutations, adaptive evolution of complex entities would be practically impossible.

In contrast to the rapid response produced by plasticity, if the production of newly favored phenotypes requires new mutations, the waiting time for such mutations can be prohibitively long and the chance of subsequent loss through drift can be high. (Pfennig et al. 2010: 459–460)

One upshot, then, is that no genetic novelty is required to initiate an 'appropriate', adaptive response to a change in one feature of an organism by other features. In these instances, genes collectively have it in their repertoires to produce the appropriate output, under previously unexperienced circumstances. The developmental systems in which they are embedded are responsible for eliciting the appropriate gene action.

5.3.2   Evolvability and gene networks

Genes play their characteristic role in development as parts of suites or networks with complex regulatory topologies (Davidson 2010; Meir et al. 2002). I take these regulatory topologies to be extragenetic in the sense that they are not intrinsic structural properties of individual genes, but highly organismed, dynamic structures.[10] Gene networks are robust, and their robustness is crucial both to the origin of evolutionary novelty and to the stability of organismal form (Ciliberti, Martin and Wagner 2007b; Wagner 2011, 2012).

Ciliberti et al. (2007b) and Wagner (2011) have modeled the dynamics of these networks extensively. Each simulation begins with initial input levels of gene products and a particular topology of regulatory relations between elements of the network. 'Mutations' and changes in initial conditions are then introduced, as either quantitative or qualitative changes in gene products, or in

---

[10] That is to say that these topographic features are not properties of the gene's 'code'.

the regulatory relations among genes in the networks. The outputs of these networks are mapped in a vast multidimensional 'network space'. Along any given axis, immediate neighbours in this space differ from one another by a single genetic mutation.

The space of viable phenotypes exhibits some remarkable features. In general, in these systems there are many more genotypes than phenotypes: '[t]his means any one phenotype typically has many genotypes that form it (Wagner 2011: 71)'. Furthermore, the space of viable phenotypes is clustered and connected. Viable networks are clustered in the sense that the genotypes that produce a particular phenotype all occupy a single small region of the overall state space. They are connected in the sense that any network capable of producing the system's characteristic phenotype can be accessed from any other through a series of single mutations without leaving the space of viable networks. Such a connected network of networks is called a 'neutral network'.

The evolutionary significance of these neutral networks cannot be overestimated. The neutral networks demonstrate that gene networks are robust. They can compensate for perturbations by producing changes in function that preserve their viable phenotype. Robustness is measured as the number (proportion) of nearest (single-mutation) neighbours in the neutral network that produce the typical phenotype. Gene networks with many viable neighbours are capable of withstanding a considerable degree of perturbations, both mutational and environmental. Wagner (2011) demonstrates the enormous capacity of such systems to maintain their stability under mutational change. For gene regulatory networks of a size typically found in biological systems, two networks producing the same phenotype may share only 20 per cent of their regulatory interactions (Wagner 2011). The constancy of form, the reliability of development and the transgenerational stability of phenotypes are secured by the adaptive robustness of their gene regulatory systems.

Developmental orchestration and the dynamics of gene regulatory networks suggest some surprising things about the relation between individual genes, and developmental systems in the joint production of phenotypes. Across a significant range of environmental and *genetic* conditions, the same phenotype is reliably produced. The constancy of the phenotype is secured by the regulative control of gene action by gene regulatory networks, and by extragenetic developmental factors. Moreover, the same genes can be made to produce a variety of stable viable phenotypes. Here again, the difference maker is the influence of the extragenetic features. This suggests, then, that whether a collection of individual genes produces a particular appropriate phenotype on an occasion depends very heavily upon the regulation of their activities by the developmental system of which they are a part.

5.3.3 Developmental instruction

With some understanding of the relation between genes, phenotypes and extragenetic developmental systems, we can take up the question of whether, generally, genes can be meaningfully understood as issuing commands for the building of phenotypes. Translated into our informational idiom, these considerations suggest that in general individual genes carry comparatively little imperative content about the production of appropriate phenotypes. Genes do not carry more information about appropriate extragenetic action, than they do about the conditions in the world in which genes operate.

Take an example: suppose a gene mutates in response to an environmental condition, say, cigarette smoke or ultraviolet radiation.[11] As these sources generally induce fairly specific kinds of lesions (Ciccia and Elledge 2010), this change in gene function carries information about the state of the world that caused it. But the nature of these mutations radically underdetermines the (appropriate) phenotypic response. The extragenetic developmental system may compensate in ways that restores the normal phenotype, or it may, as in the case of the genetic networks discussed above, produce a novel stable phenotype, or it may mount a DNA repair response, or it might initiate cell apoptosis, depending upon the kind, location and severity of the lesion. Which phenotypic output is produced is principally influenced by the response of the developmental system, which in turn depends upon a whole raft of extragenetic conditions. So, an alteration to a gene caused by a certain set of conditions – a mutation – may carry information about the state of affairs that induced it, but allows us to infer very little about the regulatory response that will eventuate.

The underdetermination of phenotype by genotype is suggestive. Our naturalised account of illocutionary content, implies that genes taken severally do not embody a signal with imperative force. Genes do not, in general, encode instructions for specific phenotypes.[12]

On the contrary, recent empirical work in developmental genetics suggests that the 'informational asymmetry' tables are turned. Developmental systems respond to circumstances in part by regulating the activities of genes in ways that produce appropriate outputs (phenotypes) across a range of circumstances. Because extragenetic developmental systems regulate gene action in such a way that they produce the same output across a range of conditions one can glean more information from them about the phenotypic outputs of gene action than one can about the conditions under which genes are acting. This suggests, then, that the regulation of gene action by extragenetic developmental systems has the informational profile of a signal with imperative force.

---

[11] These seem to be the two most common exogenous sources of DNA damage in humans (Ciccia and Elledge 2010).

[12] Even if they do encode instructions for amino acid sequences, or precursor mRNA.

If this speculation is correct, then the orthodox Modern Synthesis picture of genes exerting executive control over developmental systems in the production of phenotypes has got things more or less backwards. In order for genes to instantiate a program for building an organism, they would have to encode commands. But, in all but exceptional circumstances, this seems not to be so. Instead the control of gene action resides principally in extragenetic developmental systems. Extragenetic regulatory structures tell genes what to do. Properly construed, the 'information paradigm' works strongly against the supposition that genes generally encode a program.

Let this not be misunderstood. There are cases in which genes really can be considered to embody this kind of imperative illocutionary force. As John Maynard Smith says: 'there is no difficulty in saying that a gene carries information about adult form; an individual with the gene for achondroplasia will have short arms and legs' (Maynard Smith 2000: 189). These are cases in which the Method of Difference is appropriate for making inferences about the difference-making capacities of genes.[13] But the more we find out about the role of organisms in regulating their own development, the rarer these cases appear to be. The relation between genes and extragenetic features is one of reciprocity 'The passage of information is not simply one-way, from genes to function. There is two-way interaction . . .' (Noble 2006: 50). If so, the 'genetic program' metaphor generally misrepresents the relation between genes and extragenetic factors in the production of phenotypes.

Similar considerations apply to the idea that genes constitute a 'recipe' for building an organism (Bateson 1976; Noble 2006). Recipes are not like blueprints or computer programs, but they do, nevertheless, issue commands.[14] Genes generally don't.

5.4   From discrete genes to reactive genomes

The gene, either in its classical or molecular incarnation, began as a unit of phenotypic control. It was a 'unit', in the sense that each gene was presumed to have a relatively discrete, context independent effect on a unit of phenotype. So discrete and context independent were these effects thought to be, that it seemed natural to suppose that they would be disclosed by the kinds of investigative protocols that are appropriate to the study of well-behaved, nicely 'composing' complex causes. Genomes, for their part, were just aggregates of unit genes.

The image of genes as clear and distinct causal agents, constituting the basis of all aspects of organismic life, has become so deeply embedded in both popular and

[13] See Chapter 4.
[14] 'Recipe', after all, comes from the imperative form of the Latin '*recipere*', to take.

scientific thought that it will take far more than good intentions, diligence or conceptual critique to dislodge it. So, too, the image of a genetic program – although of more recent vintage – has now become equally embedded in our ways of thinking, along with it attendant conviction (as Jacob and Monod first put it) that 'the genome contains not only a series of blue-prints, but a co-ordinated program of protein synthesis and the means of controlling its execution'. (Keller 2000: 136)

This is a compelling, even beguiling picture, but this particular foundation stone of the Modern Synthesis is rapidly eroding. It simply doesn't stand up to empirical scrutiny.

The idea of the genome as an aggregate of genes, each of which carries some discrete parcel of information about building some distinct part of an organism, is now being set aside (Pigliucci 2010). Barnes and Dupré (2008) argue that in the last decade or so we have seen a transition from genes to genomes as the units of explanatory significance for development and inheritance. Attention has increasingly turned to the genome, as an integrated system causally embedded in its environment (Gilbert 2003b). For their part, genes perform their duties only as components of metastable, reactive complex systems – genomes (Lamm 2011).

Today's genome, rather than a set of genes initiating causal chains leading to the formation of traits, looks like an exquisitely sensitive reaction (or response) mechanism – a device for regulating the production of specific proteins in response to the constantly changing signals it receives from the environment. (Keller, 2013: 17)

Denis Noble (2012) stresses the primacy of extragenetic factors in controlling the activities of genes.

DNA sequences do absolutely nothing until they are triggered to do so by a variety of transcription factors, which turn genes on and off by binding to their regulatory sites, and various other forms of epigenetic control, including methylation of certain cytosines and interactions with the tails of the histones that form the protein backbone of the chromosomes. All of these, and the cellular, tissue and organ processes that determine when they are produced and used, 'control' the genome. (Noble 2012: 57)

Genes are beholden to organisms, cells, tissues and genomes, and are under their control. This has implications, of course, for the 'genetic program' conception of individual genes as command centres of development. It shows us that the computer program conception of the genome is erroneous: 'It is rather a "read-write" memory that can be organised in response to cellular and environmental signals' (Noble 2012: 57).[15] The shift in the rhetoric is striking. Not only are genes now 'memory', rather than 'programs', it is now cells and

---

[15] The metaphor of the genome as a 'read-write memory' also figures prominently in Shapiro (2011).

environments that send the 'signals'. Genes and genomes fall under the control of extragenetic factors. Organisms even engineer their own genomes.

Cells operate under changing conditions and are continually modifying themselves by genome inscriptions.... Research dating back to the 1930s has shown that genetic change is the result of cell-mediated processes, not simply accidents or damage to the DNA. This cell-active view of genome change applies to all scales of DNA sequence variation, from point mutations to large-scale genome rearrangements and whole genome duplications. (Shapiro 2013: 287)

The restructuring of the genome by the cell is commonplace, and important. A vivid illustration is to be found in the single-celled eukaryote *Oxyrichia trifallax*. Like most ciliates *Oxytrichia* has two nuclei – a somatic macronucleus and a germline micronucleus. During reproduction, the macronucleus of the new cell originates as a copy of the micronucleus. It then undergoes substantial reorganisation. More than 90 per cent of the germline DNA is excised, and much of the remainder rearranged (Chen et al. 2014). This is an extreme example, but it serves to underscore the way that organisms use, modify and reconstruct their genomes. The capacity of cells and organisms to alter the structure of their genomes further suggests that one of the bulwarks of the Modern Synthesis – the Weismann barrier – is no barrier at all. The genomes from which organisms are built, the genomes that participate in their daily activities, are largely of the organism's own making, and are under their control.

Reactive genomes are not isolated repositories of phenotypic commands as such. They are interactive complex adaptive systems. They exploit the resources of their environments; they regulate the capacities of their constituent genes; they buffer organisms against the vagaries of mutations and environmental perturbations. They reconcile the competing demands of the system's various parts and the exigencies of the environment. They innovate. The outcome of all this supple, goal-directed, self-synthesising activity is a stable, viable, robustly adaptive entity – the organism. Organisms, for their part, are open, ecologically embedded things that take their influences from both inner and outer sources.

A defining characteristic of a living organism is that it is an open system: there is a continual throughput of matter and free energy from the environment, and an export of entropy. Thus, a large amount of information contained in a cell derives from its biological environment and a large amount of information in an organism derives from its ecological environment ... So, genetic information is augmented by environmental information: genes may constrain the development of an organism, but they do not alone determine it. (Davies 2012: 43)

Strains of Schrödinger's 'order-from-disorder' approach to organisms are now clearly audible.

Genes provide raw materials for adaptive systems – cells and whole organisms – to use in the building and maintenance of stable, viable organisms. But they do not embody a program. If we are to persist in the use of computer metaphors, that of a database would be more apposite.

So, even at the very lowest level of the reductionist causal chain, we discover a conceptual error. The protein-coding sequences are templates. They determine which set of proteins the organism has to play with, just as a child knows which pieces of Lego or Meccano she has available for construction. Those parts of the genome are best regarded as a database. Even when we add in the regulatory and noncoding regions, there is no program in the genome in the sense that the sequences could be parsed in the way in which we would analyse a computer program to work out what it is specifying. (Noble 2012: 57)[16]

Just as genes influence biological form, biological form affects the functioning of genes.

[I]t is more accurate to think of a cell's DNA as a standing resource on which a cell can draw for survival and reproduction, a resource it can deploy in many different ways, a resource so rich as to enable it to respond to its changing environment with immense subtlety and variety. (Keller 2013: 41)

This radical reconception of the role of genes and genomes in an organism's life is not so new. It was anticipated long ago by Barbara McClintock. In her Nobel Prize lecture McClintock envisaged the genome as a reactive system that senses, monitors, regulates and co-ordinates the interactions of genes and their contexts.

In the future, attention will undoubtedly be centred on the genome, with greater appreciation of its significance as a highly sensitive organ of the cell that monitors genomic activities and corrects common errors, senses unusual and unexpected events, and responds to them, often by restructuring the genome. (McClintock 1984: 644)

Conclusion

The gene as the unit of phenotypic control has enjoyed theoretical privilege throughout the history of Modern Synthesis evolutionary biology due to its presumptive role in exerting a special, highly specific control over the development of phenotypes. Genes are said to carry information, or to embody a program or a blueprint for building an organism. This conception of gene function derives from a rather simplistic conception of gene action. On this

---

[16] Notice how easily the metaphor of the genome as 'program' is supplanted by the metaphor of the organism as agent. I shall be arguing that the latter is no mere metaphor.

view genes act as units, which influence the development of fairly discrete and specific phenotypes. Indeed this may well be true of some genes, but it is demonstrably not true of most.

Genes are not units of phenotypic control. Phenotypic control is spread throughout the entire gene/genome/organism/environment system. It is orchestrated and regulated by complex adaptive systems – cells, genomes, entire organisms – that are capable of sensing, adapting and co-ordinating responses to their conditions. These more inclusive systems – regulatory networks, genomes, cells, tissues – appear to be the loci of 'control' over the phenotype in development. The most inclusive of these control systems is the whole organism. The change in our understanding of gene function ought to occasion a reassessment of the theoretical privilege that genes have enjoyed in Modern Synthesis theorising. Here again, understanding the control of the phenotypes suggests a more central role for organisms as adaptive systems than has been customary in Modern Synthesis evolutionary thinking.

6   Fit and diversity

From competition to complementarity

There are no six-legged tetrapods. There are plenty of four-legged tetrapods, and there are many six-legged nontetrapods. One could be excused for letting this most pedestrian of biological regularities pass without a second thought. Trust Aristotle to insist that it needs an explanation.[1] He starts by positing a principle of optimal functioning: that nature works for what is best among the range of available options (Henry 2013). Aristotle argues that for organisms with an elongate body, the optimal set-up is to have two paired appendages in which in each stride the limbs must make two fixed points that describe a diagonal across the torso – first right-front to left-back, and then left-front to right-back. The mid-point of these diagonals should be more or less equidistant from the four fixed points (the feet). So convinced is Aristotle of his optimality argument that he further insists that any (blooded) organism that walks by means of more than four feet must be segmented in such a way that the segments themselves operate as more or less independent two- or four-point systems. Unsegmented organisms for which the midpoint would not be equidistant between all four feet (e.g., those with highly elongate bodies) wouldn't be able to walk effectively: 'If they had two or four they would be practically be stationary; so slow and unprofitable would their movement be (§7). They should have no legs at all.'

Aristotle's explanation of locomotion is admittedly a little obscure, but his explanatory strategy should be familiar to any practicing modern-day biologist. Aristotle is explaining of what is, why it is, and of what is not, why it is not.[2] The account of 'what is' appeals to function; the account of 'what is not' appeals to constraint. Just as Aristotle did, modern biology accounts for the fit and diversity of form in terms of function and constraint.

Among the fractionated processes of Modern Synthesis evolution, one has pride of place in the explanation of fit and diversity. Natural selection is the sole source of the adaptive bias in evolutionary change. It alone explains function.

[1] *On the Gait of Animals*, §§4–7.
[2] With apologies to Robert F. Kennedy: 'There are those who look at things the way they are, and ask why ... I dream of things that never were, and ask why not?'

In the process of promoting fit, selection also causes populations to diverge from one another. Selection thus also promotes diversity.

At the same time, it appears that natural selection might not account completely for the distribution of biological form. The reason is simple. The bearers of biological form are whole organisms. Each organism faces the tribunal of its conditions of existence as a corporate entity, not as a loose assemblage of traits. At each stage of its development from egg to adult an organism must be an integrated, functioning whole. 'This is the main difference between a frog and a Chevy. The Chevy never had to function until it was off the assembly line' (Gilbert 1999: 341). The requirement of continual viability leaves a distinctive trace on biological form. Moreover, development can only proceed with the resources at an organism's disposal at any time. Consequently, not everything that might develop could develop. So, it is to be expected that development might play some minimal role, at least, in explaining the distribution of biological form. There are, thus, two kinds of explanation of the distribution of form: selectional and developmental.

The Modern Synthesis has a well-rehearsed position on the relation of selectional and developmental explanations. As selection is fundamentally adaptive, it explains of what is, 'why it is' (at least for functional traits). As development is fundamentally conservative, it explains of what is not 'why it is not'. We have seen enough on the preceding chapters to wonder whether this particular division of explanatory labour needs to be re-evaluated.

6.1   The two-force model

Natural selection and development are often thought of as distinct, sometimes competing forces. Natural selection doesn't just winnow or destroy. It facilitates the origin and improvement of complex adaptive characters. Ernst Mayr accentuates the creative nature of selection.

When natural selection acts, step by step, to improve such a complex system as the genotype, it does not operate as a purely negative force ... It acts as a positive force that pays a premium for any contribution toward an improvement, however small. For this reason profound thinkers about evolution, such as Theodosius Dobzhansky, Julian Huxley, and G.G. Simpson, have called selection 'creative'. (Mayr 1976: 45–46)

Similarly, Francesco Ayala stresses the way that selection (and selection alone) systematically promotes adaptive evolutionary change.

Natural selection has been compared to a sieve which retains the rarely arising useful mutations and lets go the frequently arising harmful mutants. Natural selection acts in this way, but it is much more than a purely negative process, for it is able to generate novelty by increasing the probability of otherwise extremely improbable genetic combinations. Natural selection is creative in a way. (Ayala 1970: 5)

For its part, organismal development is traditionally seen as a sort of handmaiden to selection. It delivers genotypes – as phenotypes – to the arena of the shared environment, whereupon selection has at them. When development makes a discernible difference to evolution it does so only by somehow constraining or biasing the range of forms available for selection to work on (Amundson 1994; Sansom 2009). The canonical definition of a developmental constraint is given by John Maynard Smith and colleagues.

A developmental constraint is a bias in the production of variant phenotypes or a limit on phenotypic variability caused by the structure, character, composition, or dynamics of the developmental system. (Maynard Smith et al. 1985: 266)

6.1.1   Two landscapes

The putative relation between selection and development can be illustrated with two common pictorial devices: the adaptive landscape and the epigenetic landscape. Adaptive evolution is portrayed as taking place on a multidimensional surface, the adaptive landscape (Simpson 1944; Wright 1932). This surface resides in a space whose axes represent traits; one dimension for each trait. Each location in the multidimensional 'design space' thus corresponds to an individual organism's total phenotype or form. There is a further axis in addition to the trait dimensions. Each individual total phenotype has a fitness, represented as an altitude on the landscape. Individuals with higher fitness, so the story goes, generally beget phenotypically similar individuals with comparably high fitness. As evolutionary novelties are introduced into a population, some will confer yet higher fitness on their bearers. So long as the selection coefficients are sufficiently high, that is to say, the slopes are sufficiently steep, a population undergoing selection will be drawn inexorably toward a local fitness optimum. The local fitness optima are 'adaptive peaks', good locations in 'design space'. Populations at these optima are well adapted to their conditions of existence.

Adaptive evolution, then, is visualised as a process in which a population or lineage traverses a fitness surface under the influence of the force of selection. The trajectory of this process is explained exclusively (or primarily) by the topography of that surface.

Adaptive evolution is a search process – driven by mutation, recombination, and selection – on fixed or deforming fitness landscapes. An adapting population flows over the landscape under these forces. The structures of such landscapes, smooth or rugged, governs both the evolvability of populations and the sustained fitness of their members. (Kauffman 1993: 118)

The value of an adaptive topography is that it is easily visualised and so makes the evolutionary dynamics of the population intuitively clear. (Lande 1976: 315)

The way in which developmental constraint impedes adaptive evolution is sometimes illustrated by means of another pictorial device, the epigenetic landscape introduced by C.H. Waddington (1957 and 1960). As we briefly discussed in Chapter 4, Waddington represented the trajectory of a developing phenotypic feature, something he called a 'chreode', as an inclined surface.[3] The surface is marked by a series of branching channels, like river valleys and their tributaries. As development progresses, the chreode gets shunted into one channel or another, until finally it reaches its adult form.

Waddington thought that the imagery of the epigenetic landscape helped clarify some of the most puzzling features of development (Gibson and Wagner 2000). One is that there are distinct, highly specialised tissue types. Yet they all develop from the same undifferentiated, homogeneous precursors. This phenomenon is captured in the epigenetic landscape by the fact that toward the bottom of the landscape there is a limited number of channels; each is narrow and is separated from the others by steep-sided banks. Differentiation of tissues is sensitively dependent upon environmental triggers, yet it is also extremely robust. Tissues and organs develop successfully despite a wide range of environmental perturbations. This 'environmental buffering' of trajectories is represented by the banks of the valleys. A perturbation drives the developing structure up the banks of a channel, whence it returns (often enough) to the stable point at the bottom of the valley. A third feature of development illustrated by the landscape is that of genetic buffering. A population harbours an enormous amount of genetic variation but despite this variability there is startling constancy in the final products of ontogeny. Waddington coined the term 'canalisation' to cover these three salient features of ontogeny. He noted that occasionally novel, stable phenotypes may be elicited when large perturbations are applied to a canalised trajectory.

The bithorax condition of *Drosophila* is a vivid example. It appears in some individual flies where the pupae are grown in high concentrations of ether (Waddington 1956). Waddington surmised that within a population there is latent genetic variation; some individuals have the genetic endowment that permits these quite different phenotypes and some do not. But these genetic differences do not show up as phenotypic differences because development is so heavily constrained by canalisation.

Putting the adaptive and epigenetic landscapes together, we can see how developmental constraint – as represented by the canalised epigenetic landscape – might impede the power of natural selection to effect adaptive evolution, as envisaged in the adaptive landscape. Canalisation makes unavailable certain phenotypes that might otherwise be adaptively advantageous. In just the way that, for example, bithorax is adaptively advantageous in the rarefied

---

[3] See Figure 4.1, page 97.

environment of Waddington's lab, but is nevertheless unattainable for wild-type *Drosophila*. If these inaccessible areas of the adaptive landscape were attainable, then selection would be capable of driving the population onto these uninhabited peaks.

If this is how the epigenetic landscape relates to the adaptive landscape, then it bolsters the suspicion that the only contribution development can make to adaptive evolution is to constrain the adaptation-promoting effects of selection.

For instance developmental constraints frustrate selection by restricting the phenotypic variation selection has to act upon. Adaptations would be able to evolve only to optima within the constrained space of variability. (Wagner and Altenberg 1996: 973)

The nature of the existing developmental system somehow constrains or channels acceptable change [of form in evolution], so that selection is limited in what it can achieve given some starting anatomy. (Raff 1996: 294–295)

In evolution, ontogeny can only explain 'of what is not why it is not'.

This conception of the significance of development for adaptive evolution follows quite naturally from the two-force model of evolution. Selection and development are two discrete processes that exert distinct influences on the distribution of biological form. Yet only one of them systematically promotes adaptive evolutionary change. As Schwenk and Wagner put it:

This has led to a 'dichotomous approach' in which constraint is conceptually divorced from natural selection and pitted against it in a kind of evolutionary battle for dominance over the phenotype ... much of the constraint literature over the last 25 years has explicitly sought to explain evolutionary outcomes as either the result of selection or constraint. (Schwenk and Wagner 2004: 392)

There is nothing in the two-force picture itself that entails the marginalisation of development. After all, developmental constraint could prevail in the 'battle for predominance over the phenotype'. A minority position in twentieth-century evolutionary biology maintains that it does. For his part Stephen Jay Gould consistently argued that the study of evolution should largely be considered the study of how the laws of form, manifested in development, affect diversity (Gould 1977, 2002). Throughout his illustrious career, Gould tirelessly advocated for an explanatory pluralism in which the principles that govern the development of form, historical contingency and selection have a more equitable standing in evolutionary theorising.

Nevertheless, there is an implicit, yet strong presumption running through much of twentieth-century Modern Synthesis evolutionary biology that if the project of evolutionary biology is to explain the *fit* and *diversity* of organismal form, and the principal contribution of development is to stand in the way of both, then natural selection explanations must take priority, whereas development is an explanatory redoubt of last resort.

6.2   One force too many

The two-force model has two major faults: one empirical, one conceptual. The empirical error is that development does not merely constrain adaptive evolution; it is becoming evident that development makes a systematic contribution to the *adaptiveness* of evolution. The conceptual error is one we have already encountered (Chapter 2). It misrepresents the metaphysics of selection. It takes selection to be an autonomous population-level cause, rather than a higher order effect. These errors compound; taken together they conspire to misrepresent the relation between selection and development. The corrective to each deficiency requires a recasting of the role of organismal development in adaptive evolution.

6.2.1   The adaptiveness of development

The integration of development into evolutionary theory has received considerable impetus from a movement called Evolutionary Developmental Biology (evo-devo). A number of recurring themes have emerged in this body of research, each of which tends to confirm the productive role of development in adaptive evolution.[4]

Regulatory evolution   An organism has in its genome a special 'toolkit' of regulatory genes that control development by influencing the timing, or the products of expression, of *other* genes. Biologists are coming to appreciate the significance of the 'toolkit' in adaptive evolution (Carroll, Grenier and Weatherbee 2000). Regulatory genes and processes can direct the development of similar phenotypes despite differences in the underlying developing tissues and despite variations in the structural genes (Stern 2000). Additionally, changes in the regulatory role of genes can produce significant changes in the structure and function of phenotypes whose development they regulate.

Gerhard and Kirschner's (2005, 2007) conception of 'facilitated variation' illuminates the significance of regulatory evolution for adaptive change. An organism's genome consists of a core of conserved components. These are highly regular structures shared in common across organisms whose latest common ancestor must have been in the Cambrian or Pre-Cambrian. These core components underlie the production of a bewildering variety of different phenotypic structures both within organisms and between lineages. For example, the homeobox gene, *lab*, found in *Drosophila* and other insects regulates the correct development of the maxillary and mandibular segments on the head.

---

[4] These themes are nicely surveyed in Laubichler (2009).

Hox 4.2 of mammals regulates the development of the atlas/axis complex of the vertebral column. While these are wholly different structures – neither the arthropod nor the vertebrate structure has a homologue in the other group – their respective homeobox genes, *lab* and *Hox 4.2*, are startlingly similar in their structure (Duboule and Dollé 1989). Other arthropod and vertebrate homeobox pairs are 'virtual base-pair for base-pair copies' of one another (McGinnes et al. 1990).

These highly conserved core structures and processes combine with more variable developmental resources in the production of novel phenotypes. The preserved core-variable periphery structure of developmental systems confers on organisms a dynamic robustness, the ability to maintain viability through mounting compensatory changes to perturbations. It also confers on development the capacity to 'search' morphological space, alighting upon new, stable and adaptive structures. This suggests an enhanced positive role for development in promoting adaptive evolution.

> The burden of creativity in evolution, down to minute details, does not rest on selection alone. Through its ancient repertoire of core processes, the current phenotype of the animal determines the kind, amount and viability of phenotypic variation the animal can produce ... the range of possible anatomical and physiological relations is enormous .... (Gerhard and Kirschner 2007: 8588)

Sean Carroll and colleagues identify two major consequences of the co-option of highly 'conserved' regulatory networks for adaptive evolution.

> First, conserved regulatory circuits can be recruited for new roles during the development of novel morphologies ... In this way, large numbers of genes may be deployed in a novel structure with just a small number of regulatory changes. Second, evolutionary changes in gene regulation can facilitate morphological diversification of a novel character. As regulatory evolution modifies the genetic interactions within a developmental program, new patterns can emerge both within and between species. (Carroll et al. 2000, 158–159)

The redeployment of old regulatory resources in new ways seems to be the predominant means of evolutionary change. It certainly appears to facilitate adaptive evolution.

> Most evolutionary change in the metazoa since the Cambrian has come not from changes of the core processes themselves, but from regulatory changes affecting the deployment of the core processes ... Because of these regulatory changes, the core processes are used in new combinations and amounts at new times and places. (Kirschner and Gerhard 2005: 221–222)

Innovation   Because development is dynamically robust, it confers on organisms the capacity to innovate. We have already seen some of the implications of this. Through its phenotypic repertoire, and the attendant phenotypic accommodation to new forms, developmental processes are capable of producing new stable forms, *without* the need for genetic mutations (Moczek et al. 2011; West-Eberhard 2003). 'The recurring theme ... is the *creative role played by* evolutionary changes in gene regulation' (Carroll et al. 2000: 167, emphasis in original). Wagner (2014) stresses the way that the structure of developmental architecture confers on development a property he calls 'innovability', the biased tendency to produce stable, novel structures.

Modularity   Developmental systems have these marvellous capacities largely because of their modularity (Bolker 2000). Developmental modules are internally tightly integrated units. They have weak regulatory interactions with one another. This is a general feature of adaptive, self-organising systems:

In principle, a modular system allows the generation of diverse phenotypes by the rearrangement of its internal module connections and relatively independent evolution of each module .... (Kitano 2004: 830)

The fact that each module is tightly integrated internally, and strongly decoupled from others, secures two significant features of developmental dynamics. Internal integration ensures that each module is robust. It can compensate for perturbations. Decoupling ensures that a module is protected against potentially disruptive changes occurring in other modules. Typically, each module is capable of producing any number of a large array of stable outputs (Bergman and Siegal 2002; Greenspan 2001; Von Dassow et al. 2000). Which of its capacities is manifested on a particular occasion is determined by the context in which the module finds itself. Taken together, these properties of modular architecture confer on an organism the capacity robustly and reliably to produce a viable organism typical of its kind, and to generate phenotypic novelties as an adaptive response to genetic, epigenetic or environmental perturbations.

Plasticity   Adaptive phenotypic plasticity appears to be a prerequisite for adaptive evolution (Pfennig et al. 2010). Phenotypic plasticity makes development responsive to its circumstances.

Responsive phenotype structure is the primary source of novel phenotypes. And it matters little from a developmental point of view whether the recurrent change we call a phenotypic novelty is induced by a mutation or by a factor in the environment. (West-Eberhard 2003: 503)

Recent work by Andreas Wagner (2011, 2012, 2014) vividly underscores the way that phenotypic robustness drives adaptive phenotypic evolution. We encountered Wagner's work in the preceding chapter. There we used it to show

how the architecture of regulatory networks helps to secure the constancy of form required by inheritance. Here we use it to highlight the way in which development makes a systematic contribution to the adaptive bias in evolution.

Wagner plots the way that gene networks robustly produce phenotypes. Neutral networks have remarkable adaptation-promoting properties; they act as evolutionary capacitors. Moreover, neutral networks for different phenotypes overlap in gene space. This means that different phenotypes can be produced by the same, or very similar, gene networks. One neutral network may overlap with many others. This in turn allows development to 'search' the space of viable phenotypes efficiently. Wagner concludes that the robustness of development, as manifest in the structure of gene networks, systematically contributes to adaptive evolution in two ways:

First, robustness causes the existence of genotype networks, complex web-like structures formed by genotypes with the same phenotype, which facilitate phenotypic variability. Second, a robust phenotype can help the evolutionary exploration of new phenotypes ... by accelerating the dynamics of change in an evolving population. (Wagner 2012: 1256)

The integration of ontogeny into evolutionary theory that has been inaugurated by these new disciplines urges a wholesale reconsideration of the traditional role in which organismal development has been cast.

Evo–devo argues that the variational capacities of genomes are functions of the developmental systems in which they are embedded, for example, through their modular organization, the dynamics of their mechanistic interactions and their nonprogrammed physical properties. (Müller 2007: 946)

It appears that development contributes to adaptive evolution in three ways (Hendrikse, Parson, and Hallgrímsson 2007; Pfennig et al. 2010):

(i) It biases the direction of variation. This bias is not wholly independent of adaptation as development regularly produces adaptive novelties.

(ii) It regulates the amount of phenotypic variation. The buffering of development against the effects of mutations serves to secure the viable production of phenotypes. There are many more genotypes than phenotypes, thanks to the regulatory power of development.

(iii) It serves as an evolutionary capacitor, storing up latent variation, and phenotypic repertoire that influences the rate and direction of evolution in the future.

These considerations urge a shift in emphasis away from natural selection, and toward ontogeny, as the cause of adaptations.

The contribution of development to evolution places considerable strain upon the traditional Modern Synthesis conception of the relation between selection and development.

The explanation of adaptive change as a population-dynamic event was the central goal of the Modern Synthesis. By contrast, evo–devo seeks to explain phenotypic change through the alterations in developmental mechanisms (the physical interactions among genes, cells and tissues), whether they are adaptive or not. (Müller 2007: 945–946)

Evo-devo's exponents are clear about the profound implications:

... an inclusion of developmental systems properties into evolutionary theory represents a shift of explanatory emphasis from the external factors of natural selection to the internal dynamics of developmental systems .... (Müller and Newman 2005: 489)

The message is clear: development must be included in our picture of how organisms come to be suited to their environments (Schlichting and Moczeck 2010). If this is correct, then it is inappropriate to think of natural selection as the only source of adaptive bias in evolutionary change. It is equally misguided to think of the contribution of development as restricted to a constraint on adaptive evolutionary change. That should raise some concern, as it is practically an axiom of the Modern Synthesis evolution that only selection causes adaptively biased evolutionary change. The relation between selection and development needs to be rethought.

5.2.2   The nature of selection

In light of this need, the second, conceptual error of the two-force model becomes particularly pertinent. The traditional Modern Synthesis picture radically misrepresents the nature of selection in a way that conspires against a genuine recognition of the productive role played by development in evolution.

The two-force model sets selection and development in competition with one another, because it casts them both as processes that independently cause changes in population structure. For any given population change, then, it would appear reasonable to ask how much of the change is due to selection and how much to development. Or at least it makes sense to ask how much the population would have changed were selection permitted to act unimpeded by developmental constraint. But we have already encountered reasons to suppose that such a question is nonsensical. In Chapter 2, selection was characterised as a higher order effect. It is a reflection, at the level of population structure, of the aggregate of the several causes of individual-level births, deaths and reproductions.[5]

Selection is spontaneous, in the sense that nothing needs to be added to the collection of individual-level causes of birth, survival, reproduction and death, to get a population to undergo selective change. To suppose that there is a

---

[5] This is not to insist that selection is *not* a cause of population change. It may be (although I happen to think it isn't), but that issue is not at stake here. See Walsh (2007a, 2010a, 2013a).

population-level cause – selection – that is *additional* to the causes of individual births, deaths and reproductions would be to overpopulate the world with causes. In this sense, selection is causally redundant (Brunnander 2007).

That is not to say that selection is *explanatorily* redundant. As we also discussed in Chapter 2, there is a special class of explanations that appeals to the dynamics of ensembles. I called them 'higher order effect explanations'. These explanations demonstrate the way that an ensemble-level trend or pattern emerges from the several, complicated, sometimes apparently disordered activities of entities in the ensemble. Gases move toward thermodynamic equilibrium, substances diffuse, mists fall and populations undergo adaptive change.

The two-force picture, then, errs in casting natural selection and development as alternative, independent causes of population change. They are related, of course, but not as opposing forces, but as part to whole. Selection, at the population level, is composed of development (broadly construed) at the individual level, in just the way that the pressure of a volume of gas is composed of the kinetic energies of its molecules. Selection is the higher order effect of individual-level causes of population change. Development is the individual-level cause of population dynamics. They are not, in any robust way, ontologically independent. Here the lesson we drew in Chapter 2 from Darwin's discovery, that the activities of individual organisms are sufficient to explain fit and diversity, becomes highly germane. All the causes of adaptive evolution are to be found at the level of individual organismal development broadly construed. Therefore, all the causes of bias in evolution are to be found in development.

That, in a nutshell, is the conceptual error of the two-force picture. If adaptive evolution *just is* development, it is misguided to suppose, as the orthodox Modern Synthesis view does, that development is principally *opposed* to adaptive evolution.

6.3 The causes of adaptation

How can populations of organisms undergo adaptive evolution? The orthodox Modern Synthesis theory offers an elegant answer. Adaptive evolution occurs through the accretion of small, rare, independent, beneficial mutations. Selection eliminates the deleterious ones and retains and combines the good. And the consequence is well-adapted organisms. This story may well be right, in its way. But it underspecifies what conditions need to be met in order for mutations to accumulate in this way.

How can such wonderful systems emerge merely through random mutation and selection? For if Darwin told us that adaptation occurs through the gradual accumulation of useful mutations, he has not yet told us what kinds of systems are capable of accumulating useful mutations. (Kauffman 1993: 173)

The traditional Modern Synthesis approach, as we discussed in Chapter 2, locates the sources of adaptation in the intrinsic properties of genes. It posits genes (or replicators) as independently acting units of small phenotypic effect, whose activities are reasonably constant across contexts. When genes behave in this way, the adaptiveness of organisms arise gradually from the gradual accretion of 'good' genes.

Increasingly, an alternative is taking form. It is becoming apparent that adaptive evolution demands certain specific features of individual organisms. The investigation of the features of organisms required for adaptive evolution falls under the rubric 'evolvability' (Sansom 2011). We can think of evolvability as posing the question: 'what features must organisms possess in order that lineages of them can evolve?' (Kauffman 1993; Kirschner and Gerhard 1998; Von Dassow and Munro 1999; Wagner and Altenberg 1996).[6] The significance of evolvability can be glimpsed in the fact that some lineages of organisms are much more susceptible to adaptive evolution than others.

Lineages contrast in the *actual pattern* of their evolutionary histories: some are strikingly more disparate ... than other apparently comparable lineages. In some cases it is plausible to suppose that these differences reflect differences in *evolutionary* potential rather than reflecting chance or selective environment. There is a further temptation to think that these differences in evolutionary potential themselves have an explanation in the developmental biology of the organisms concerned: lineages differ at a time and over time in *evolvability*. (Sterelny 2007: 164, emphasis in original)

For all this tells us, the dynamics of individual development may merely facilitate adaptive diversity. Yet, it is evident that development does much more. Certain specific features of development are a prerequisite for the adaptiveness of adaptive evolution. The kinds of systems that are capable of accumulating 'useful' mutations are those that can balance two apparently antagonistic demands – stability and mutability. Schwenk and Wagner point out the paradoxical requirements of evolvability.

On the one hand, phenotypes must be mutable and therefore responsive to the constantly changing demands of the environment ... On the other hand, phenotypes must be stable so that the complex dynamics of their developmental and functional systems are not disrupted. (Schwenk and Wagner 2004: 390)

---

[6] Sterelny (2007), Sansom (2009) and Brown (2013) present a variety of views on the nature of evolvability.

It first became apparent that the conditions that promote evolvability are far from trivial when computer programmers attempted to exploit the principles of evolution in order to 'evolve' improved computer programs, by introducing random mutations to a code, and then selecting the best resulting program. It turns out that not just any computer program can evolve in this way. Most programs will crash if random mutations are introduced. Programmers quickly found out that in order for such a program to undergo adaptive improvement it must have very special architectural features (Wagner and Altenberg 1996).

An evolvable program must be structured in such a way that it is not adversely affected by most changes, but also that a few changes will move it to a new, stable state. The key to this capacity is modular architecture. This is a phenomenon we have already encountered. For all intents and purposes, the evolvable computer programs must be modular in the same way that organismal development is. Systems with this sort of architecture manifest the kind of robustness that is so distinctly characteristic of organisms. This suggests that the capacity of a population to undergo adaptive evolution is grounded in the capacities of organisms, particularly those surveyed above that are manifested in their development: modularity, plasticity, robustness, innovation. These are all features of the adaptiveness of individual organisms. Moreover, each of these capacities is in its own right, a kind of developmental constraint: 'a bias in the production of variant phenotypes or a limit on phenotypic variability caused by the structure, character, composition, or dynamics of the developmental system' (Maynard Smith et al. 1985: 266). The traditional Modern Synthesis picture of the role of constraint in adaptive evolution thus appears to be profoundly mistaken. Constraints on the development of individual organisms do not *impede* adaptive evolution. They are the principal causes of adaptive evolution.

6.4   Complementarity

The competition for explanatory relevance engendered by the two-force model is predicated upon the supposition that *both* selectional and developmental explanations are mechanistic. They compete, on the traditional picture, because they both cite discrete, alternative mechanisms of population change or stasis. But selectional and developmental explanations of evolution, properly understood, do not compete; they are complementary.

Noncompetition is a consequence of a special feature of explanations. Explaining is more than merely describing events; it involves capturing counterfactually robust patterns (Garfinkel 1981). Explanations explain by identifying relations of counterfactual dependence. By this I mean (minimally) that to explain a phenomenon, e, is to cite some feature of the world, c, such that were c to occur, e would as well. Any given event may instantiate many different counterfactual dependences, each of which may be suitable to be used in an explanation. So, any given event may be susceptible to multiple, complete and independent explanations.

6.4.1   Diffusion

The diffusion analogy that we borrowed from Schrödinger in Chapter 2 nicely illustrates the multiplicity of independent explanations. Any episode of diffusion may be explained in either of two ways – by adverting to either the 'arrangement' of molecules in solution, or their 'distribution'. The arrangement can be specified by giving a compendious list of each molecule and its location and velocity. The distribution consists of the variation – the unevenness – in velocities in different parts of the solution. Arrangement and distribution are different properties. A specification of the arrangement does not need to mention the distribution. Conversely, a specification of the distribution does not need to mention any of the nonrelational properties of any individual.

The arrangement explanation will cite the way that each of the properties of each of the molecules at one time – their locations, momenta and velocities – and the interactions between molecules caused the arrangement at a later time. It tells us that this instance of diffusion occurred in this solution because this molecule moved in this direction, and that molecule moved in that direction and this other molecule collided with yet another. To be sure, it is a little cumbersome, but it is an explanation nevertheless. The distributional explanation cites the way that the distribution of velocities at one time leads to the distribution of velocities at the later. It tells us that diffusion occurred because of a spontaneous tendency of a solution to move from an uneven to an even (equilibrium) distribution of velocities.[7]

These are noncompeting, complementary explanations of the same phenomenon: an episode of diffusion. They appeal to different properties of the solution in their explananda, and they take different properties of the solution as their respective explanantia. The first – 'arrangement' explanation – is an 'individual-level' explanation. It cites and explains properties of individuals. The second – 'distribution explanation' – is an 'ensemble-level' explanation it cites and explains properties of the ensemble.

Furthermore, they have different implications, and different applications. The individual-level explanation shows the ensemble-level effect to be causally necessitated by the activities of the particles, severally. It entails that were this collection of individual motions to occur again, the very same higher level effect would also occur. The distributional explanation, for its part, tells us that

---

[7] Galton's explanation of the pattern of inheritance (Chapter 2) is another, classic example of a distributional explanation (see Ariew, Rice and Rohwer 2015).

this particular *pattern* and *rate* of diffusion does not depend upon any *particular* specific arrangement of individual velocities for its particles. The rate and direction of diffusion depend only upon the *distribution* – the unevenness – of velocities in the solution. The rate and direction of diffusion are, in fact, largely *independent* of the particular details of the particles' motions, because the counterfactual dependence between the starting distribution and the final distribution does not depend upon any particular arrangement. We should expect that similar diffusion phenomena should occur across a wide range of systems, with different arrangements of velocities amongst its component molecules. The same change in distribution will occur robustly even with different chemical solutions. Because 'arrangement' and 'distribution' explanations account for different features of a system, and have different modal implications, they are different explanations. Because they can be applied to the same event, they are complementary.

The complementarity of individual-level and ensemble-level explanations transposes to the evolutionary case. Developmental explanations tell us how the individual-level causes operating within an ensemble conspire to produce the higher order effect – selection. Distributional – natural selection – explanations tell us how the distributional properties of the ensemble at one time are a function of their distributional properties at a previous time. These are noncompeting, complementary explanations of the same phenomena.

6.4.2   The two-level model[8]

This all points to the need to replace the two-force model. I suggest a two-level model. We have seen that explaining adaptive evolution – fit and diversity – involves two distinct projects: (i) explaining changes in *population* structure and (ii) explaining the adaptedness of *individuals*. Natural selection theory accomplishes the first of these: populations undergo changes as a function of their structure, the *distribution* of individual-level causes of living, dying and reproducing. But it does not explain the properties of individual organisms, so it does not accomplish the second project. The second project, explaining the adaptedness of individuals, requires an account of the processes occurring within individuals that dispose them to preserve and initiate adaptively advantageous phenotypes.

My suggestion is that developmental biology offers the prospect of just that: a theory of the causes of adaptive evolution *within individuals*. In particular, developmental constraint confers on individuals the kind of stability and mutability required for the maintenance and initiation of adaptive phenotypes. A theory of development gives us a major part of what Marjorie Grene (1961)

---

[8] This subsection relies heavily on the discussion in Walsh (2003).

calls 'a causal study' of adaptive evolution. All the causes of adaptive evolution are to be found at the level of individual organisms, particularly in the dynamics of their development.

A developmental theory of the individual-level causes of adaptive evolution would not by itself explain why biological populations in general tend to undergo predictable changes in their trait structure. This is precisely the role that our population-level theory of natural selection plays. The theory allows us to abstract away from the specific arrangement of individual-level causes of change within a given population, and to fix our attention on the *distribution* of these properties. These distributional properties allow us to project evolutionary changes from one specific arrangement of individual-level causes to others, and to generalise across all biological populations.

The two-level approach does not cast natural selection and development in competition for explanatory relevance. Each is (potentially) complete in its own distinct domain. Nothing about individual-level causes needs to be added to natural selection theory in order to get it to explain why populations change in their trait structure. Similarly, nothing about the structure of populations would need to be added to a complete theory of adaptive development in order for it to explain its proprietary phenomena, *viz.* that organisms whose developmental processes constrain form in various ways preserve and initiate adaptive phenotypes.

These two explanations of evolution are complementary. Each casts adaptive evolutionary population change as an instance of a different kind of counterfactually robust regularity: one regularity depends upon the individual-level causes, the other on the distribution of trait fitnesses.

While there is a degree of explanatory *independence*, between developmental and selectional explanations, there is also a considerable measure of ontological *dependence*. The distributional properties that figure in selectional explanations are fixed by the individual-level features of organisms that are realised in their development (broadly construed). The explanandum of the natural selection explanations – changes in trait structure – are realised in the lives and deaths (most particularly, the development) of individual organisms.

In this respect, a theory of development and a theory of natural selection are related, as are Classical Mechanics and the statistical treatment thermodynamics. The first theory in each pair identifies forces operating at the level of the components of an ensemble. The second explains ensemble-level phenomena that are the consequence of its structure.

Given a population of molecules in a gas, Newtonian mechanics explains why each individual molecule behaves in the way it does. But the individual-level mechanical forces operating within a given ensemble of molecules do not explain what all such ensembles have in common, by dint of which they behave in predictable ways. For that we need the theory of thermodynamics. Analogously, given a population of organisms, the mechanics of development explain why individual organisms tend to be adapted to their conditions of existence, tend to produce adaptive innovations. Natural selection, for its part, explains why populations comprising these kinds of individuals should change in predictable ways.

Conclusion

Modern Synthesis evolutionary theory prioritises natural selection in the explanation of the distribution of biological form. Selection alone promotes, and hence explains, fit and diversity. The explanatory primacy of selection implies a minimal role for development. Development, according to Modern Synthesis orthodoxy, is fundamentally constraining. Therefore, it can only really explain of what is not, why it is not. This picture comprehensively misrepresents the relation between organismal development and natural selection. There are two sources of error, empirical and conceptual. The traditional picture underestimates the many ways that organismal development introduces a systematic bias into adaptive evolution. However, empirical studies in developmental biology are progressively rectifying this oversight. The conceptual error misconstrues the relationship between development and selection. It misrepresents selection as a discrete cause of adaptive population change. Rather, selection, properly understood, is an aggregative, higher order effect of all the processes occurring within and to individual organisms. It adds no causes that aren't already accounted for by the suite of causal processes occurring within individuals. The source of the adaptive bias in evolution, then, must reside in these individual level processes. Evolution is adaptive because organismal development is adaptive.

Here again, an article of Modern Synthesis doctrine comes under threat. Fractionation, as we saw, holds that development plays no role in imparting to evolution its adaptive bias. So, the argument goes, something else must. Here the Modern Synthesis must posit a wholly independent process, natural selection, to do the job. But this separation of development and adaptation, and the concomitant separation of development from selection, are highly suspect, on both empirical and conceptual grounds. In the next chapter we explore the proper place of development in adaptive evolution.

7   Integrating development

Three grades of ontogenetic commitment

At best estimate, at any one time an adult human comprises 3.72×10^{13} cells (Bianconi et al. 2013).[1] That's a big number – at a couple of orders of magnitude larger than the estimated number of stars in the Milky Way (Howell 2014), 'astronomical' barely covers it. Yet a human being, any multicellular organism for that matter, is no massive nebulous of cells. Our cells, more than 200 different types of them, are exquisitely organised into integrated functional suites of highly specialised organs and tissues. In the process of making an organism, our cells regulate, alter, relocate, induce, produce, terminate and communicate with each other. All this prodigious, orchestrated complexity originates from a single cell. The collection of processes that intermediate between this evidently inauspicious start and its startling results – 'the differentiation of cells, ... The collective behavior of cells in the formation of tissues and organs, ..., and growth, or increase in mass ...' (Trinkaus 1984: 2) are known as ontogeny (organismal development).

It hasn't gone unnoticed that ontogeny is important. In fact, throughout the history of comparative biology, theories of the diversity of form have tended to be ontogenetic in character.[2] The motivating thought running through much of the history of biology seems to have been that to explain the distribution of biological form, one must know the developmental processes by which biological forms are produced.

The study of ontogeny attained an historical high water mark around the time that evolutionary thinking began seriously to take hold. But evolution did not eclipse development – certainly not initially, anyway. The two were mutually supporting. From its inception, evolutionary thinking was well poised to· assimilate the insights of developmental biologists. The relevance of ontogeny for evolution was well-known to Darwin himself. He repeatedly cited the inspiration that he drew from the embryological works of von Baer, for example.[3] The nineteenth-century embryologists – Wilhelm Roux, August

---

[1] That's just counting the human cells!

[2] Needham's (1934) magisterial history of developmental biology traces thinking about development from Classical Indian thought through to the early twentieth century.

[3] Much to the latter's dismay (Oppenheimer 1967).

Weisman and Ernst von Haeckel and their colleagues – adopted and extended Darwin's evolutionary views. Well into the twentieth-century embryologists and evolutionists – such as E.S. Russell (1945), Walter Garstang (1922), Gavin De Beer (1938), I.I. Schmalhausen (1948 [1986]), C.H. Waddington (1957) and J.T. Bonner (1958) – sought to integrate the study of ontogeny into evolutionary thinking.

Yet, developmental biology seems to have been the principal casualty of the fractionation of evolution ushered in by the Modern Synthesis theory of evolution. Viktor Hamburger (1980) famously noted that development had been more or less omitted from the Modern Synthesis as it emerged in the early twentieth century, and proceeded to subsume the various biological subdisciplines – systematics, paleontology, ethology, ecology – under its ambit. Development, Hamburger argues, was treated as a 'black box', a process necessary for organismal evolution, yet one whose details make little difference to our understanding of evolutionary dynamics. It was not supposed to be so. In announcing the inauguration of the Modern Synthesis, Julian Huxley proclaimed that ' ... a study of the effects of genes during development is as essential for an understanding of evolution as are the study of mutation and that of selection' (Huxley 1942: 41).[4] All the same, as evolutionary biology progressed, the study of development became increasingly peripheral. By the later decades of the century, Bruce Wallace's suggestion that '... problems concerned with the orderly development of the individual are unrelated to those of the evolution of organisms through time ...' (Wallace 1986: 149) was not out of keeping with the tenor of the time.

Towards the end of the twentieth century, however, the disunity of evolution and development began to appear unsustainable, to some biologists at least. The growth in understanding of the role of genes in development, the increased emphasis on the influence of developmental timing on form (Gould 1977), the discovery of a highly conserved genetic 'toolkit', all pointed to the need for a re-integration of ontogeny into evolutionary thinking. Developmental biologists increasingly began to argue that development and evolution are '... neither mutually exclusive, nor under independent control' (Hall 1999: xvi).

A concerted movement to promote the importance of development for the process of evolution seems to have taken root in a conference in Dahlem, Berlin, in 1981 (Love 2015). Its objective was '... to examine how changes in the course of development can alter the course of evolution and to examine how evolutionary processes mold development' (Bonner 1982: frontispiece). The Dahlem Workshop is credited with sowing the seeds of Evolutionary Developmental Biology that is currently enjoying such a spectacular

---

[4] I first encountered this passage in Morange (2011).

efflorescence. In drawing together researchers from such disparate fields as molecular and cell biology, life history evolution, and the dynamics of macro-evolution, the workshop established a framework for synthesising research in developmental biology and evolutionary biology. Indeed, many of the themes that have been avidly pursued in the interim years – the control of development by genes, the role of the timing of development, extragenetic control of gene action, the relation of selection, adaptation and constraint, the role of developmental plasticity, the nonrandomness of novelty – were first articulated at this workshop.[5]

Bonner's introduction to the Dahlem proceedings is a veritable fanfare of optimism. Recounting a chance meeting at an airport with a customs official who enquires whether he has heard of von Baer, Bonner recalls: 'This convinced me of something I had already suspected: we had chosen exactly the right moment for a conference on 'Evolution and Development' (Bonner 1982: 15). The enthusiasm generated by the Dahlem conference was contagious. It inspired startling advances in the molecular biology and cytology of development, the place of heterochrony in evolution, the significance of robustness and plasticity. Yet, thirty-five years on from Dahlem, the place of development in evolution is still uncertain. Despite repeated calls for the re-evaluation of evolutionary thinking in the light of empirical advances in the understanding of development (Pigliucci 2009a, 2009b; Pigliucci and Müller 2010), there is little general appreciation of how development might impact evolutionary theory.

In this chapter we survey various strategies for integrating organismal development into evolutionary thinking. I call them 'three grades' of ontogenetic commitment to reflect the fact that they represent a hierarchy of increasingly intimate involvement of ontogeny in evolution. The process of going through the grades culminates in a position I call 'Developmental Holism' in which organismal ontogeny is the central unifying concept in inheritance and in adaptive population change, quite some distance from the marginalisation that Hamburger (1980) decried. Developmental Holism not only places ontogeny at the centre of evolution, it also envisages a radically reconfigured place for natural selection in adaptive evolution, a place foreshadowed in the preceding chapter. The distinguishing feature of Developmental Holism is the role it extends to the agency of organisms. Organisms are purposive systems. This fact is abundantly evident in their development, and crucially implicated in evolution.

7.1   Three grades of ontogenetic commitment

We can map out various conceptions of the place of development in evolution on a scale of increasing influence of ontogeny on evolution.

---

[5] Alan Love's (2015) collection of essays traces the influence of Dahlem on current evo-devo.

7.1.1   Grade 1: Development as constraint

We encountered Grade 1 ontogenetic involvement in the preceding chapter. Its objective is to preserve the structure of the Modern Synthesis, while incorporating the obvious fact that development contributes to evolution. Grade 1 is predicated on the supposition that development is fundamentally conservative (Lewens 2009a), and does not participate in the processes of inheritance, nor does it introduce evolutionary novelties. It simply constrains the amount of variation available for selection to work on (Pearce 2011).[6]

Grade 1 is the traditional Modern Synthesis view. It continues to hold a significant amount of currency. Even now, it is taken as a truism that selection and constraint constitute two opposing forces in evolution. A recent *Nature* article begins: 'Evolution involves interplay between selection and developmental constraint' (de Bakker et al. 2013: 445). These authors surmise that the differential pattern of digit loss in the Archosauria represents an evolutionary trade-off. On the one hand (so to speak) the retention of digits may reflect the conserved activities of genes active in the early development of the limb bud. On the other, digit loss is promoted by '... selection pressures for limb functions such as flying and perching' (de Bakker et al. 2013: 445).

To be sure, development can constrain evolutionary change, but we have already encountered ample reasons to abandon the idea that that is all it can do. Development does more than just apply the brakes. In Chapters 4 and 5 we saw that development also contributes to inheritance, there construed as the pattern of transgenerational phenotypic resemblances and differences. Likewise, in Chapter 6 it became apparent that organismal development is creative and not merely conservative, and therefore contributes positively to the adaptive bias in population change. So clearly Grade 1 involvement underestimates the significance of development for evolution. The question is what to replace it with.

7.1.2   Grade 2: Development and selection

One common approach is to suppose that the Modern Synthesis conception of evolution can take on these features of development without undergoing any particularly momentous revisions (Sterelny 2000b). For example, while the plasticity manifest in development is increasingly recognised as a primitive feature of living things (Schlichting 2003), and its involvement in evolution has been widely championed, its contribution has been thought by some to be wholly in keeping with Modern Synthesis separation of development from inheritance and selection. Plasticity contributes to adaptation in a variety of

---

[6] See Amundson (1994) for a classification of kinds of developmental constraints.

ways, by amplifying novel changes, by buffering organisms, by obscuring and then exposing, the variability in a population available for selection to 'work on'. Moreover, it can facilitate adaptive evolution by 'finessing' the problem of the orchestration required for the evolution of complex adaptations, through phenotypic accommodation.

> Phenotypic accommodation finesses the problem of correlated change: a genetically-caused modification in one system need not wait for a genetically-caused change in associated systems, even when both must change for either to be adaptive. (Sterelny 2009: 99)

Nevertheless, its capacity to contribute to evolution is also under the control of – is explained by – natural selection.

> On this view, adaptation is a real phenomenon, and is rightly explained by selection, but it is made possible only by very special developmental scaffolding. The bedrock agenda of evolutionary biology is to provide an explanation of the evolution of that scaffolding. (Sterelny 2000b: S385)

This is an appealing line of thought. While it integrates the dynamics of development into the study of adaptation, it also preserves a distinctive place for selection enshrined in the standard interpretation of Modern Synthesis evolution. This proposal envisages a mere minor adjustment to Modern Synthesis thinking, and it appears prominently throughout the evolutionary literature in various forms and strengths. Some authors (e.g., Debat and David 2001; Gibson and Wagner 2000; Schwenk and Wagner 2001; Wagner 2000) suppose that selection and developmental processes both play a significant role in explaining adaptive evolution, but that ultimately the relevant features of development are themselves consequences of selective forces. That is to say, plasticity and the adaptiveness of development are implicated in evolution and accounted for in the way that any other adaptation is treated, as an advantageous trait to be explained, like any other advantageous trait, exclusively in terms of the creative power of selection (Godfrey-Smith 1996).

Other proposed revisions are more sensitive to the primitive contribution that the adaptiveness of individual ontogeny makes to adaptive evolution. Kauffman (1993), as we saw in Chapter 2 for example, argues that adaptive evolution takes place only on a certain kind of adaptive landscape, one with multiple peaks and relatively shallow slopes. The topology of the adaptive landscape, in turn, is determined by the dynamics of individual development. Populations of individuals that are robustly buffered against a range of perturbations, and are capable of producing adaptive innovations in response to other perturbations, inhabit adaptive landscapes propitious for evolution. So the conditions required for a *population* to undergo adaptive evolution are secured by the developmental dynamics of *individuals* that compose the population.

These individual organisms occupy the 'boundary region' between chaotic dynamics and rigid, imperturbable dynamics. Kauffman further speculates that 'natural selection may be the force which pulls complex adaptive systems into [the] boundary region' (Kauffman 1993: 219), where the generation and maintenance of stable phenotypes through self-organisation is possible. Other authors contend that the features of development exert the strongest force on the distribution of form, whereas selective forces are weak (Goodwin 1994).

Like the attempts to assimilate development in Grade 1, these proposed second grade revisions are also inadequate, if only on account of being incomplete. For one thing, they countenance no role for ontogenetic development in inheritance. In fact some prominent sponsors of these revisions explicitly retain the traditional Modern Synthesis conception of inheritance as grounded exclusively in genes. There are no new evolutionary characters that are not inaugurated by changes in genes. We already noted Sterelny, for example, saying of phenotypic novelties produced by the plasticity of development that '[s]uch novelties have no effects on the germline are not inherited [*sic*]' (Sterelny 2009: 94).

Others, however, are decidedly more sanguine about allowing that developmental processes can initiate evolutionary – that is to say, inheritable – characters: 'Responsive phenotypic structure is the primary source of novel phenotypes. And it matters little from a developmental point of view whether the recurrent change we call a phenotypic novelty is induced by a mutation or by a factor in the environment' (West-Eberhard 2003: 503).

Developmental Systems Theory (DST) is a further, more comprehensive attempt to promote the standing of ontogeny in evolution. 'DST' is an umbrella term encompassing a range of views concerning the relation between development, inheritance and evolution (Oyama 1985; Oyama, Griffiths and Gray 2001). One of the principal features of DST is its laudable insistence that the causal responsibility for the production of phenotypes is spread throughout the entire gene/organism/environment system. Epigenetic, morphological, ecological, behavioural and cultural influences all contribute to development in such a way that the control of the phenotype cannot justifiably be said to reside exclusively, or in any privileged way, in any particular developmental resource. All of these causes of form can initiate new stable forms, and all typically are required to maintain their transgenerational stability.

DST sets itself up as a challenge to the *ontology* of the Modern Synthesis. What is inherited, according to DST, are not traits *per se*, much less genes, but '... any resource that is reliably present in successive generations, and is part of the explanation of why each generation resembles the last' (Griffiths and Gray 2001: 196). In general, development cannot be explained by decomposing the respective contributions of its component causes. 'The outcomes of development are explained at the systems level, and developmental is influenced by the context in which it unfolds, leading to an extensive conception of that system' (Griffiths and Tabery 2014: 89). To this extent DST is strongly in line with the role of development in inheritance that I outlined and endorsed in preceding chapters.[7] Furthermore, it holds that what *evolves* are entire developmental systems. 'Fundamentally, the unit of both development and evolution is the developmental system, the entire matrix of interactants involved in a life cycle' (Griffiths and Gray 2001: 206). In these respects DST provides a welcome corrective to some of the excesses of the orthodox interpretation of the Modern Synthesis.

Nevertheless, it is not entirely clear that either the attempts to insinuate development into evolutionary dynamics (briefly surveyed above) or DST, serves adequately to attribute to development the status that it really should have. These proposals all fall short in two crucial ways. First, in failing to recognise the proper relation between development and natural selection in adaptive population change – the relation discussed in the preceding chapter – they tend to perpetuate the erroneous conception of natural selection as an autonomous cause of adaptive change, over and above organismal inheritance and development. Griffiths and Gray's (1994, 2001, 2005) treatments of selection and adaptation illustrates the traditional role that DST retains for selection as the cause of adaptive bias in evolution: '... advocates of DST who have discussed its implications for evolutionary theory have stressed the importance of selection in designing developmental systems that can assemble the same matrix of developmental resources in each generation ...' (Griffiths and Gray 2005: 419).

Second, while these proposals commendably emphasise the extended – 'causally spread' – character of development and inheritance, they typically decline to identify the privileged role of the organisms in these processes. They do not explicitly, at least, accord a place to the agency of organisms in the explanation of inheritance and development. DST does not explicitly recognise the purposive goal-directedness of organism as the locus of control and regulation of these processes. Indeed, its exponents advertise the fact that DST is committed to the pursuit of methodological mechanism (Griffiths and Tabery 2014).

These amendments to the Modern Synthesis occupy a middle ground – a second grade of ontogenetic commitment. They acknowledge that development contributes more to adaptive evolution than merely imposing a check on the process of selection. In this, they all represent a significant advance on Grade 1. Yet, they also seek to maintain the Modern Synthesis conception of the relation between development and selection. Some versions even hypostatise

---

[7] Indeed, I'm happy to acknowledge the influence of DST on the formulation of my own views. I thank Susan Oyama for helpful discussions on this matter.

multiple, independent processes, or 'levels', of selection (Pigliucci 2009a, 2009b). I shall dub this cluster of positions 'Grade 2 ontogenetic commitment'. It may well lay claim to being the newly emerging orthodox position on the relation between development and adaptation (Laland, Odling-Smee and Feldman 2014; Pigliucci 2009a, 2009b; Pigliucci and Muller 2010).

7.1.3   Grade 3: Developmental holism

There are two motivating ideas behind the third grade of ontogenetic commitment. The first is the claim canvassed in Chapters 2 and 6 that selection is the higher order effect on a population of the activities of individual organisms. The second is that the organisms are purposive entities. These two theses are closely related. It is because organisms are purposive, goal-directed systems of a certain kind, that there is no need to think of selection as a discrete cause that introduces adaptive bias into population change.

The important lesson from DST, and related studies of ontogenetic development, is that the proper development of organisms involves an enormously complex and widely spread system of causes. But that should not be taken to imply that there is no unit that exerts executive control over development. Whereas in the orthodox Modern Synthesis that unit is the gene, in the envisaged alternative it is the organism as a purposive entity. Proper development depends upon the capacity of organisms to assimilate, integrate and orchestrate the causal contributions from genes, epigenetic structures, tissues, organs, behaviour and the physical, ecological and cultural setting.

An organism is a self-regulating, self-forming system.[8] It is a goal-directed entity that transduces causal influences from internal and external sources into an organised, integrated whole. It regulates, assimilates and organises those inputs into a system capable of pursuing the goals that constitute its particular way of life. For the purposes of capturing these phenomena, the primary unit of explanatory significance is not the gene, nor is it the developmental system – interpreted as the complete set of causes of form – rather it is the organism as purposive system, as the locus of control and regulation of these causes.

Echoing Kant, we can say that organisms are the locus of control because they build themselves. They synthesise the very materials out of which they are made. There is a reciprocity between the causal contributions of an organism's several parts and the activities of the system as a whole. Just as the activities of the system as a whole are the causal consequence of the activities of the component parts, so too the activities of the component parts are controlled and regulated by the system as a whole. Such systems are characterised by

---

[8] In fact, an organism is more like a maximal aggregate of such units. Each organism will have among its parts other self-building, self-maintaining, self-regulating systems.

a form of causal holism of the sort we explored in Chapter 4. Their causal architectures feature extensive feedback loops constructed in such a way that the causal contribution of any one part percolates through the system, and then depending on the contributions of other parts, redounds upon itself. The causal components of the system manifest what Wagner (1999) calls 'nonseparability'.

In a nonseparable system, each behaviour is *jointly* caused by the complete suite of its components. We cannot generally attribute specific effects, or specific differences in effect, to specific components of the system. Moreover, the parts of the system do not have context insensitive effects. A change in the activity of one part changes the activities of all others. The consequence of this causal holism is that every phenotype of every organism, novel or inherited, is jointly the effect of causes both internal to and external to the organism. These causal influences are commingled in such a way that one cannot partition the effects on form into those caused by the environment external to the organism, and those caused by factors internal to the organism.

Insofar as any single entity can be said to 'control', 'regulate' or 'orchestrate' this widely distributed plexus of causes, it is the organism as a whole. As a self-organising, self-synthesising, self-regulating, purposive system, the organism has the capacity to co-ordinate and marshal these causal influences toward the attainment of a stable, viable, adaptive system.

Grade 3 involvement takes seriously the two-level view of the relation between development and adaptive evolution adumbrated in Chapter 6. All the causes of adaptive evolution take place at the level of organismal struggle: living, reproducing and dying. The adaptive bias in evolution is introduced by the adaptiveness of organisms. On Grade 3 ontogenetic commitment, ontogeny is the central unifying phenomenon in evolution. The adaptive plasticity of organisms manifest in their development causes (i) the reliable recurrence of form from generation to generation, (ii) the origin of adaptive novelties and (iii) the increasing adaptedness of organisms in a population. Adaptive evolution is caused by the adaptiveness of organismal development. Indeed evolution – inheritance, novelty, adaptive population change – is development writ large. As the embryologist Walter Garstang remarked: 'Ontogeny doesn't recapitulate phylogeny: it creates it' (Garstang 1922: 98). Grade 3 commitment augments Garstang's dictum with the adage that ontogeny creates adaptive evolutionary change, too.

By locating the cause of the adaptive bias in evolution in the adaptive activities of organisms, particularly in their development, the Grade 3 inter-pretation does not need to invoke natural selection to do the job. On this account, natural selection is what happens to a population of organisms, and when it does, it realises adaptive population change precisely because of the adaptive bias that organismal development introduces into the dynamics of the population.

Because organisms are adaptive systems that generate adaptively novel traits and maintain them across generations, there is no need to posit a supra-organismal process, distinct from the activities of organisms to promote adaptive population change. Adaptive population change is spontaneous given the purposive activities of organisms – it 'follows inevitably', in Darwin's happy phrase.

Of course, much more needs to be done to establish the case that organismal purposiveness underlies the contribution of development to adaptive evolution. As we discussed in Chapter 1, modern evolutionary biology has foresworn natural purpose. It operates in an 'etiolated' world of mechanisms and mechanistic explanations. A considerable amount of methodological and metaphysical ground-clearing needs to be undertaken in order to carve out a role for organismal purposiveness in evolution. These tasks are taken up in earnest in Chapters 8 through 11. In the meantime, however, it is worth considering the implications of Grade 3 commitment for the fractionation of evolution, on which the entire edifice of Modern Synthesis evolution is built.

7.2   The end of fractionation

The great boon of fractionation was the idea that the component processes of evolution could be observed and studied independently of one another. Each of these processes has its own distinctive cause. One doesn't particularly have to know much about development in order to study the mechanism of inheritance, the origination of novelties, or adaptive evolutionary change. They are reasonably independent of one another, and none of them depends particularly intimately upon development.

Fractionation is the bold empirical wager of the Modern Synthesis. If we cannot properly study the processes of evolution in isolation from one another, then we cannot properly understand evolution as a loose aggregate of the quasi-independent components. This is an issue to which empirical findings in biology can be addressed. It is on these empirical grounds that the Modern Synthesis should stand or fall.

7.2.1   Fractionation and Grade 3 commitment

Over the course of the last four chapters we have charted the way that the inclusion of organisms as active participants in evolution has undermined the central convictions of the Modern Synthesis, particularly fractionation.

In Chapter 4 we saw that the separation of inheritance and development is theoretically unmotivated and empirically unwarranted. The observations that support inheritance holism suggest that the *pattern* of inheritance – the intergenerational stability of interlineage differences and intralineage resemblances – is upheld by more than just the transmission of genes. The processes of development, broadly construed, are intimately involved. There is generally no way to hive off the contribution of genes to the pattern of inheritance from any other of the influences on form. From the perspective of Grade 3 ontogenetic commitment, the presumed independence of inheritance and development is an unsubstantiated article of Modern Synthesis dogma, that is neither borne out empirically, nor justified conceptually. The Modern Synthesis view of inheritance as comprising exclusively the transmission of replicated entities is unfounded.

Grade 3 involvement also undermines the various putative distinctions between inherited and noninherited characters. Inheritance, under Grade 3, is considered to be the *pattern* of resemblances, rather than the process of transmission. The pattern that is secured by development admits of degrees. Some phenotypes are extremely robust, recurring reliably generation on generation, across an enormous range of conditions. Some are much less so. It is to be expected, then, that inherited/acquired is not a genuine biological distinction (Keller 2011b; Oyama 1985), but rather at best a continuum. Nevertheless, according to Grade 3 commitment, any transgenerationally stable trait – of any degree of constancy – is potentially an evolutionary trait. Any such trait is capable of contributing to differential organismal survival and reproduction, and might increase or decrease in frequency on account of doing so. Likewise, Grade 3 acknowledges a variety of processes capable of initiating evolutionary novelties, from genetic mutation, to developmental plasticity, to environmental change. The initial impetus for these novel traits may differ, but each is produced by the entire developmental system, incorporating its interactions with genes, genomes, cells, tissues, developmental modules and environments.

Finally, Grade 3 commitment denies the Modern Synthesis distinction between development and selection as discrete causes of form. By locating the cause of the adaptive bias in evolution in the adaptive activities of organisms, particularly in their development, the Grade 3 interpretation does not need to invoke natural selection to do the job. Selection is the aggregate of individual development (broadly construed), death and reproduction. The adaptive bias in form is *manifest* in populations, but it originates in organismal development.

Fractionation is the flagship commitment of Modern Synthesis evolutionary theory. It is the presumed quasi-independence of the component processes of evolution that has facilitated the startling growth of gene-centred evolutionary thinking, and at the same time, it this supposition, more than any other, that has marginalised organisms from evolutionary theory. What we are beginning to find out about inheritance, development, novelty and adaptive change suggests that the component processes of evolution cannot be considered in isolation from one another. These findings also promote a conception of evolution that is more sensitive to the contribution of organisms. If Grade 3 ontogenetic commitment is the correct way to accommodate the contributions of organisms to evolution, it is difficult to see how the fractionation of evolution, so crucial to the Modern Synthesis, might survive.

7.2.2   The myth of two spaces

Grade 3 ontogenetic commitment strongly challenges the Modern Synthesis metaphor of two spaces and a barrier. That idea was that evolutionary processes could be compartmentalised into those that take place exclusively in phenotype space and those that take place exclusively in genotype space. Inheritance, mutation and recombination are 'genotype space' processes. Natural selection, the interactions of organisms with their environments, and with one another, learning and behaviour occur in phenotype space. There is, further, a process of 'mapping' from genotype space into phenotype space, organismal development. There is also a barrier, the Weismann barrier, preventing the transmission of changes that occur within organisms in phenotype space back into genotype space. In this picture, only those processes that register as changes in genotype space count as genuinely evolutionary. The only processes that occur in phenotype space that can induce changes in genotype space are those that involve organisms leaving or entering the population: selection, drift, immigration and emigration. All other phenotype space processes are non-evolutionary processes.

   The two spaces and a barrier picture of evolution is ill-conceived. In Grade 3 involvement, the putative distinction between exclusively 'genotype space' processes, and 'phenotype space' processes dissolves. The processes of 'genotype space' are inextricably entwined with the processes of 'phenotype space' in a way that renders the distinction between them nugatory:

[T]he genotype–phenotype concept that is currently in wide use within evolutionary theory conceals the fact that it is an abstraction of a relation that is the outcome of very complex dynamics that in many cases are intimately connected to the environment. (Noble et al. 2014)

Moreover, the holism of development and inheritance renders the very idea of a barrier between phenotype space and genotype space an irrelevance. Even if there is no way in general to write developmental, behavioural and cultural changes that occur within an organism into its genes. But that in no way means that these innovations are not inheritable, or that the changes they induce are not evolutionary.

There is not a domain of inheritance set off against an entirely distinct domain in which behaviour, learning, social interactions (and the like) are played out. Nor is there any bounded entity in which the information required for building an organism resides. The generation of organisms, the initiation of evolutionary novelties, the inheritance of features, these all arise from the interaction of complex, reactive, adaptive organisms with the influences of their environments and their constituent genes, in such a way that the causal contributions of genes, genomes, cells, tissues and environments cannot generally be disentangled. There is, then, no distinction in kind between 'ecological' events and 'evolutionary' events, as Sterelny (2009) suggests. Any event may be either or both.

With the collapse of fractionation, the two-spaces and a barrier metaphor becomes completely unmotivated, and in fact misleading. The dissolution of fractionation, and the obsolescence of the 'two spaces' metaphor, invites a reformulation of evolutionary thinking, one in which the purposive activities of organisms begin to displace the gene as the central unifying concept in evolution.

Conclusion

Considerations adduced in the preceding four chapters recommend a comprehensive revision of Modern Synthesis evolutionary thinking. That account of evolution marginalised organisms. But this diminished role for organisms no longer seems viable. Empirical advances in the understanding of development, suggest that development should be seen as the unifying phenomenon in evolution. Development, broadly construed, is the process that holds in place the pattern of inheritance required of evolution, it is the source of adaptive bias in evolution, and the origin of evolutionary novelties. Giving development its due requires ascending to what I have called the 'third grade of ontogenetic commitment'.

In Grade 3 commitment, development is the manifestation of the purposiveness of organisms. In development organisms orchestrate, integrate, accommodate and negotiate the various causal influences from genes, genomes, epigenetic factors, cells, tissues and environments in the production of a stable, highly adaptive responsive entity. That, in turn, requires acknowledging the significance of organismal purposiveness for evolution. This is the issue we turn to now.

Part III

Situated Darwinism

The most acute defects of the Modern Synthesis issue from its marginalisation of organisms and its excessive reliance upon genes as the canonical unit of evolution. Organisms are different from genes. Genes, on the Modern Synthesis view, are mechanico-computational devices, encoding information, and acting as a centralised command centre for the construction of organisms. Organisms are purposive, self-synthesising, self-regulating entities, open systems, constantly exchanging matter and energy with their environments. Part II charted the various ways in which gene-centred thinking may be inadequate for the purposes of understanding evolution. Along the way, it made admittedly vague gestures toward the ways in which evolutionary thinking might be transformed by placing the purposive activities of organisms at its core.

In this section, I say something a little more positive about the prospect of another way of thinking about evolution. I call this nascent alternative 'Situated Darwinism', for reasons that I hope will become evident as we proceed. The crucial difference between evolution construed in this way and the Modern Synthesis approach is that, according to the latter, evolution is fundamentally a molecular phenomenon. According to Situated Darwinism, evolution is fundamentally an ecological phenomenon. By 'ecological phenomenon' I mean that evolution arises out of the engagement of purposive entities – organisms – with their affordances. *Affordance* is a special theoretical concept, borrowed from ecological psychology, to be spelled out in more detail as we proceed. Roughly speaking, an affordance is a joint property of a purposive system and the conditions with which it interacts. Affordances are opportunities for, or impediments to, the pursuit of a system's goals. Affordances thus imply agency. Only agents experience their conditions as affordances, and conversely, conditions can only afford opportunities or impediments to agents. In this way, the alternative I shall outline introduces the organism as agent into evolutionary thinking.

In order to engage with an affordance, an agent must have two features. First, it must be able to experience its conditions *as affordances*. That is to say that it must generally be capable of responding to propitious conditions *as propitious* by exploiting them, and to unpropitious *as unpropitious*, by ameliorating them.

Second – and concomitantly – a system must also have an adaptive repertoire. That is to say that on any occasion, there must be a range of possible outcomes or activities that the system or its parts could implement. Which elements of the system's repertoire are actualised on an occasion must generally be biased in favour of those that are conducive to the attainment of the system's goals. It is here that the goal-directed plasticity of organisms becomes integrated into adaptive evolution. The adaptive plasticity of development is a manifestation of organisms' ability to enlist their phenotypic repertoires in response to, and in alteration of, their affordances.

Just as Modern Synthesis thinking is propelled by its own methodological precepts, so is the alternative. Situated Darwinism is driven by methodological commitments and metaphysical assumptions of its own. They include: organisms as natural agents, the commingling of form and affordance, the appeal to unreduced teleology. I am aware that these notions are perplexing to some, and a provocation to others, and so I take some time to spell them, and their implications, out more fully.

Chapter 8 addresses the issue of adaptation as seen from the perspective of Situated Darwinism. Adaptation is not the process in which the external environment moulds passive form. Rather it is the process by which organisms respond to, and in the process create, their own system of affordances. This is a significant departure from the Modern Synthesis conception of adaptation, and it will need some explaining.

Affordances entail purposes. But the very idea of purpose in nature is rebarbative to many scientists and philosophers of a naturalistic bent. So, if the idea that evolution is the consequence of organisms' engagement with their affordances is to gain any credence, we shall have to rehabilitate the notion of a natural purpose.

Chapter 9 attempts to do so. It argues that purposes are a part of the natural world. This is in no way mystery mongering, or dabbling in the occult. Nature has purposes because purposes, or goals, are just the end-states achieved and maintained by goal-directed systems. Goal-directedness, for its part, is an observable property of a system's dynamics. In naturalising purposes, Chapter 9 also develops the outline of an account of the way that purposes might figure in genuine scientific explanations. These explanations are non-causal, and teleological: an event or occurrence is explained by appeal to the goal that it subserves. Teleological explanations do not supersede mechanistic explanations, nor do they reduce to them. They are autonomous from, and complementary to, their mechanistic counterparts. A satisfactory evolutionary biology must find a place for both.

Chapter 10 argues that a theory that takes the purposive contribution of organisms to evolution seriously would have to be a completely different kind of theory from the Modern Synthesis. It would have to be what I call an 'agent' theory. An agent theory represents the dynamics of a system under study as initiated by an agent in response to conditions that are partially constituted by the agent itself. In contrast, the Modern Synthesis is an object theory. It represents organisms as objects affected by extraneous initial conditions and external causes. If the process of evolution arises from the engagement of agents (organisms) with their affordances, then the Modern Synthesis is incapable of accurately capturing it.

Chapter 11 attempts to draw the various strands of the emerging alternative conception of evolution into a cohesive whole. It lays out the ways in which the central concepts of evolution are reconfigured under Situated Darwinism. Specifically, it argues that the battery of distinctions that are so crucial to Modern Synthesis thinking disappear when viewed through the lens of Situated Darwinism. There is no difference in kind between inherited and acquired characters, between evolutionary and nonevolutionary population change, between the processes that conserve form and those that alter it. Moreover, it argues that the Modern Synthesis presumption of the distinctness of inheritance, development, selection and the production of novelties are ill-conceived.

8   Adaptation
Environments and affordances

Ridiculing one's opponents' favourite mode of explanation seems to have been a popular Enlightenment contact sport. Molière treated us to his sham doctor of the dormitive virtues, and almost a hundred years later we meet Dr Pangloss. He is Voltaire's parody of Leibniz. Pangloss's (Leibnizian) conviction that God would have no reason to create anything other than the perfect world constrains him to explain every natural phenomenon on the presumption that 'all is for the best in the best of all possible worlds'. The inane optimism that ensues even has him putting the best possible 'gloss' on the devastating Lisbon earthquake.[1]

Stephen Jay Gould and Richard Lewontin (1979) take Dr Pangloss as the inspiration for their own parody of the optimistic excesses of what they call the 'Adaptationist Programme', in their landmark paper *The Spandrels of San Marco*.[2] As they see it, the adaptationist programme '... regards natural selection as so powerful, and the constraints upon it so few, that direct production of adaptation through its operation becomes the primary cause of nearly all organic form, function and behaviour ...' (Gould and Lewontin 1979: 584–585). The epithet they apply to adaptationism, 'The Panglossian Paradigm', signifies that it blithely assumes every organism to be optimally adapted to its conditions of existence.

It isn't very surprising that Gould and Lewontin's adaptationist parody has done little to derail the 'Adaptationist Programme'. Adaptationism isn't absurd or ridiculous; in fact it has been a very sober and successful research programme, an integral part of the growth of Modern Synthesis evolutionary biology. Generations of adaptationists since Gould and Lewontin's challenge have simply denied the charges. Biological form might be less than ideally adapted to its conditions of existence or, at best, only some elements of form may be ideally adapted. Some concede that while selection is the most powerful force operating on a population, it is important not to neglect other factors. Other adaptationists seek to evade the charge of 'metaphysical Panglossianism',

---

[1] '... all that is for the best. If there is a volcano at Lisbon it cannot be elsewhere.'

[2] *The Spandrels of San Marco: A Critique of the Panglossian Paradigm* is a classic. By most recent count it has received 5,518 citations (Google.scholar.ca).

maintaining that the perfect adaptedness of biological form is simply a heuristic, or a 'methodological stance' (Dennett 1995), or a starting point for investigation (Lewens 2009b).[3] Insofar as adaptationism deserves any censure, many concede, it is due to a predilection for excess Panglossianism among a few of its adherents. As long as this proclivity is held in check, the adaptationist programme is beyond reproach (Orzack and Sober 1994).

It is unfortunate that the dispute about adaptationism within evolutionary biology and its philosophy should have crystalised around Gould and Lewontin's allegation of Panglossianism. It has rather waylaid the real debate, diverting attention away from another, much more nuanced challenge to the Modern Synthesis approach to adaptation woven through *The Spandrels* (and related works).[4] We can make out the contours of the critique, and the vague outline of a positive alternative to the adaptationist programme in the antepenultimate and final sentences of *The Spandrels*. The former encapsulates the critique, and the latter gestures toward the alternative.

Too often, the adaptationist programme gave us an evolutionary biology of parts and genes, but not of organisms.... A pluralistic view could put organisms with all their recalcitrant yet intelligible complexity back into evolutionary theory.[5] (Gould and Lewontin 1979: 597)

In the preceding chapter we caught a first glimpse at some of the implications of putting organisms with 'all their recalcitrant complexity' into evolutionary thinking.[6] In this chapter, I build on the groundwork laid there. In the process, I shall seek to articulate in more precise detail this inchoate alternative to the Modern Synthesis conception of adaptation gestured toward in *The Spandrels*. In a nutshell, adaptive evolution arises from organisms' purposive interactions with their conditions of existence.

8.1   Adaptationism

Adaptationism falls directly out as a consequence of the fractionated evolution of the Modern Synthesis. As we discussed in Chapter 6, when none of the

---

[3] Godfrey Smith (2001a) and Lewens (2009a) offer slightly differing taxonomies of the kinds of adaptationism.

[4] The related works in question include Lewontin (1978, 1983, 2001).

[5] For the sake of completeness, the omitted sentence is: 'It assumed that all transitions could occur step by step and underrated the importance of integrated developmental blocks and pervasive constraints of history and architecture.'

[6] Huneman (2010) surveys some of the various calls to reintroduce organisms into evolutionary thinking, and the challenges they raise. Talbot (2013) reckons that recent work in molecular and developmental genetics of the sort we surveyed in preceding chapters is fulfilling Gould and Lewotin's prophecy: 'Given the revelations now pouring forth from the world's molecular biological laboratories, ... the organism is being given back to us as we have always known it – whole, full of surprises, ...' (Talbot 2013: 68).

processes occurring within organisms introduces an adaptive bias to form, something else must. Doing so is the unique domain of natural selection. It moulds form to meet the challenges posed by the external environment. That account of selection, in turn, incurs two adaptationist commitments: the autonomy of the environment and explanatory externalism.

8.1.1   Environmental autonomy

The exigencies of the organism's environment determine which organisms survive and reproduce. By the process of promoting some organisms, and eliminating others, the environment selects, winnows and changes biological form. The result is a population comprising those individuals best suited to surviving in those very environmental conditions. Insofar as a trait is an adaptation, then, it must be identifiable as a response to pressures exerted on form by an autonomous environment.

[O]rganisms respond to the environment, but the environment is largely autonomous with respect to the organisms. The environment is seen as either stable (as far as the time scale of the evolutionary process in question is concerned) or else as changing according to its own intrinsic dynamics. (Godfrey-Smith 2001b: 254)

The *niche* is an apposite metaphor for the nature of organism/environment relations, as conceived by the adaptationist programme. A (nonmetaphorical) niche is a space, or a recess, into which something (say, a statue) might fit. Like a real niche, an evolutionary niche is a set of properties of an environment, to which an organismal form may fit. The features of the evolutionary niche can be specified independently of the organisms that occupy them, to the extent that we can make sense of actual niches that are empty or unoccupied.

William A. Calder III nicely illustrates this use of the niche concept in his explanation of the distinctive mode of life of New Zealand's three kiwi species (*Apteryx spp*).

I prefer to look on this curious bird as a classic example of convergent evolution. In this view an avian organism has acquired a remarkable set of characteristics that we generally associate not with birds but with mammals ... When there were no mammals present to lay claim to the niches in this hospitable environment, birds were free to do so. (Calder 1978: 142)

Ironically perhaps, Calder's use of the traditional niche concept appears in the very issue of *Scientific American* in which Lewontin (1978) first questions its coherence. A few pages later Lewontin encapsulates the concept of adaptation implicit in Calder's scenario.

The modern view of adaptation is that the external world sets certain 'problems' that organisms 'solve', and that evolution by means of natural selection is the mechanism for creating these solutions ... The concept of adaptation presupposes a pre-existing world that poses a problem to which an adaptation is the solution ... (Lewontin 1978: 213)

Organismal form and the niches to which it adapts are thus decoupled and asymmetrically dependent. Organismal form depends on niches, but niches do not depend upon organismal form.

8.1.2   Explanatory externalism

The adaptationist programme accords explanatory primacy to the adaptation promoting influences of the external environment over the presumptively adaptation-neutral processes of development and inheritance. That in turn requires that we are able to bracket off the contributions that the environment makes to evolutionary change (selection) from those that are internal to biological form (inheritance, development, mutation). Insofar as a trait is an adaptation, it is a response to pressures exerted on organismal form by the niche or external environment. To explain the manifest fact that biological form is largely well adapted to its conditions of existence we must invoke the environment external to the organism. The adaptationist programme is thus committed to 'explanatory externalism' (Godfrey-Smith 1996).

The conceptual decoupling of form and environment is supported by the same assumptions that gave us the fractionation of evolution. The process that introduces adaptive bias in evolution is distinct from, and independent of, the processes that occur within organisms. It is because we can differentiate the contributions of the environment that we can see it as the principal cause of adaptive change.

Decoupling promotes the imagery of evolution occurring in two quasi-autonomous domains. There is the inner realm of the replicator, in which replicators compete, strike up alliances, copy themselves and implement their developmental programmes. Set off against the inner realm, there is the outer realm of the niche or environment that selects, winnows and moulds biological form. One process, and one process alone – selection – fits organisms to environments.

The decoupling schema leaves scant place for organisms in adaptive evolution. Organisms are simply middlemen, the interface between these two self-standing domains.

In this view the organism is the object of evolutionary forces, the passive nexus of independent external and internal forces, one generating 'problems' at random with respect to the organism, the other generating 'solutions' at random with respect to the environment. (Lewontin 2001a: 47)

All in all, this is a powerful, compelling picture of evolution, and appears to follow straightforwardly from Modern Synthesis evolutionary thinking. It is this picture that Gould and Lewontin are challenging by advocating 'putting organisms with all their ... complexity back into evolutionary theory'.

8.2   Organism and environment

It is not entirely clear that the decoupling of organism and environment induced by the adaptationist programme is adequate to account for the adaptedness of form. One reason to suppose that it isn't is that what *does* explain adaptation is not the autonomous features of the environment, but something that intimately involves the organism itself.

Consider an example: the adaptive solution to 'the problem' of locomotion in water. Paramecia and porpoises have both solved it, but in very different ways. The differences are due to the way that water is experienced by organisms of different sizes. A harbour porpoise experiences water in much the way we do; for a porpoise water flows easily. A porpoise displaces water by setting up smooth laminar flow across its body. Porpoises have evolved a terete shape, a strong, powerful fluke, and a narrower muscular caudal peduncle to concentrate the propulsive power of the tail stroke, as adaptations to the problem of locomotion in water.

At a length of approximately 200 microns, a paramecium experiences the viscosity of water differently, much as we would experience being immersed in corn syrup (Purcell 1977). A paramecium cannot displace water by setting up laminar flow. Instead it possesses helical bands of cilia, whose rhythmic beating serves to 'screw' the organism through its thick medium.[7] These are two radically different 'design solutions' to the same environmental feature, the viscosity of the water. Yet what makes them adaptive is determined not by the physical properties of the water, but the way in which those properties are *experienced* by the organism. Cilia are an adaptive response only to the way that paramecia experience the viscosity of the water, and a caudal peduncle is a response only to the way that a porpoise experiences it. A caudal peduncle on a paramecium would be about as adaptive as cilia on a porpoise.[8]

Adaptations, then, are not differential responses to an environment *per se*, but differential responses to an *experienced environment*. Consequently, environments *per se* are insufficient to explain adaptive changes in form. The experienced environment is not external to the organism, nor is it autonomous. It consists not just in the physical features of the environment, but the way that

---

[7] A *Paramecium* actually has three 'gaits', only two of which involve the asymmetric beating of cilia. See Hamel et al. (2011).

[8] I chose an example that sounds cilia on porpoise.

organisms are affected by those features. That, in turn, depends upon the capacities of organisms. The experienced environment is determined as much by the properties and capacities of the organism as it is by the physical properties of the environment. That being so, the traditional Modern Synthesis idea that adaptive evolution requires the concept of an external environment to mould form is mistaken.

Here Gould and Lewontin's call to incorporate organisms 'with all their recalcitrant ... complexity' is pivotal. The general idea is that we should resist thinking of organisms as somehow separate from, passive with regard to, the conditions under which they evolve. Organisms partly constitute the conditions in which they evolve.

In realising this, as in so much else, C.H. Waddington seems to have been ahead of his time.

[O]rganism and environment are not two separable things, each having its own characteristic in its own right, which come together with as little essential inter-relation as a sieve and a shovel of pebbles thrown onto it. (1957: 189)[9]

Lewontin and the ecological psychologist J.J. Gibson, at virtually the same time, elaborate on Waddington's inchoate notion of the 'coming together' of organisms and environment. The confluence of views is striking.[10]

There is no organism without an environment, but there is no environment without an organism. There is a physical world outside of organisms and that world undergoes certain transformations that are autonomous ... But the physical world is not an environment, only the circumstances from which environments can be made. (Lewontin 1978)

[I]t is often neglected that the words animal and environment make an inseparable pair. Each term implies the other. No animal could exist without an environment surrounding it. Equally, although not so obviously, an environment implies an animal (or at least an organism) to be surrounded ... the environment is ambient for a living object in a different way from the way that a set of objects is ambient for a physical object. (Gibson 1979: 3)

Waddington, Lewontin and Gibson all point to the participatory role of organisms in making the conditions to which biological form adapts and evolves. This insight forms the core of the organism-centred alternative to adaptationism.

8.3   Form and affordance

The organism isn't just a middleman, caught between the machinations of the inner computational realm and the vicissitudes of the external

---

[9] I thank Dick Vane-Wright for directing me to this passage, and for helpful discussion on these issues. See Vane-Wright (2014).

[10] The parallel between Lewontin and Gibson is important for my purposes. I claim that Gibson's concept of an affordance provides a solution to Lewontin's problem of adaptation.

environment. It is a participant in evolution. Organisms are actively involved in creating and constituting the conditions of their existence. We need an alternative conception of the relation between the organisms and the conditions to which forms evolve that accommodates this insight. John Haugeland's (1998) terms *intimacy* or *commingling* work nicely for our purposes. Haugeland's notion of commingling arises from his attempt to articulate an alternative to the prominent Cartesian conception of mind. On that view, the mind is wholly internal to the agent, the environment wholly external to mind, and they communicate through the narrow channels of perception (environment-to-mind) and action (mind-to-environment).

Haugeland follows Heidegger and Merleau-Ponty in thinking of the mind not as separated from, and passively representing the external environment, but as a capacity of an agent engaged in the world, actively constituting the conditions to which it responds.

The contrary of this separation ... is something I would like to call *intimacy* of the mind's embodiment and embeddedness in the world. The term 'intimacy' is meant to suggest more than just necessary interrelation or interdependence but a kind of *commingling* or *integralness* of mind, body and world – that is to undermine their very distinctness. (Haugeland 1998: 208)

In like manner, we should think of the organism as commingled with – both making and responding to – the conditions of its existence.

8.3.1   Affordances

By their adaptive activities organisms partially constitute their experienced environments. Organisms, as reactive, purposive entities make and regulate both the conditions in which they live and how the conditions impinge on them. The conditions in which form evolves are a joint project of the organism and its setting. The leading idea is captured nicely in J.J. Gibson's concept of an affordance.[11]

The affordances of the environment are what it offers the animal, what it *provides* or *furnishes*, for good or ill ... I mean by it something that refers to both the environment and the animal ... It implies the complementarity of the animal and the environment. (Gibson 1979:127)

---

[11] Not everyone shares my enthusiasm for affordances. Robert Richards, for instance, has his doubts: 'Gibson, in his own funny way, was a realist more naive than any Medieval Aristotelian. And I think most philosophers who try to make something of his views are indulging in a kind of necrophilia of defunct science' (Callebaut 1993: 360). I thank Werner Callebaut for drawing my attention to the complaint.

An affordance is a relational, reciprocal kind of thing. What an organism's conditions 'provide' or 'furnish' is jointly determined by the extra-organismal properties of the organism's setting and the capacities of the organism itself.

We can identify an adaptation as a response to a challenge faced by the organism only once we understand how the features to which form adapts are *experienced* by the organism 'for good or ill'. One salutary suggestion, then, is that an adaptation is not so much a response to a niche or an environment, traditionally construed, but to an affordance.[12]

'Affordance' is a metaphysically rich notion. In fact, the nature of affordances is a matter of some dispute (Turvey 1992). On one appealing construal, an affordance is an emergent property of an organism/environment system (Chemero 2003; Stoffregen 2003). Affordances imply purposive systems, and conversely, purposive systems imply affordances. For a system to experience its conditions of existence *as* affordances it must generally be capable of responding to them *as affordances*, by exploiting the opportunities they provide for the attainment of its goals, or by mitigating the impediments. In turn, to be a purposive system is to be a system that is capable of responding to its conditions in ways that are conducive to the fulfilment of, or ameliorative of the impediments to, those goals. 'Affordances are opportunities for action; they are properties of the animal–environment system that determine what can be done' (Stoffregen 2003: 124). Affordances are thus properties of organism/environment systems that have 'meaning' or significance for the organism.

> The theory of affordances is a radical departure from existing theories of value and meaning. It begins with a new definition of what value and meaning are. The perceiving of an affordance is not a process of perceiving value-free physical objects to which meaning is somehow added … it is a process of perceiving a value-rich ecological object. (Sanders 1993: 290)

Affordances have special importance for adaptive evolution. Because the capacities of biological form and affordances are coconstituting, any change in one is a change in the other. Form and the affordances to which it evolves change together. Furthermore, as affordances are reflections of purposiveness, the adaptive goal-directedness of organisms structures and conditions the affordances on which evolution occurs. I discuss these implications of affordances for evolution in turn.

8.3.1   The coevolution of form and affordance[13]

The relation between form and affordance is very *unlike* the relation between form and the environment as envisaged in traditional Modern Synthesis

---

[12] This approach to organism/environment relations owes much to the philosophers Maurice Merleau-Ponty (see Matthews (2002)) and Hans Jonas (1966).

[13] This section draws heavily upon Walsh (2013a).

evolution, or for that matter in Grade 2 ontogenetic involvement. An organism's affordances are not detached from the properties of form in the ways that its physical setting might be thought to be. Nor do affordances have their 'own intrinsic dynamics' (Godfrey-Smith 2001b) in the way that the organism's environment is thought to have. The relation between an organism and its affordances is reciprocally constituting, and reciprocally effecting. *Affordance* is a Janus-faced concept, comprising both organism's environment, and the range of things it can do with its environment. Biological form and its affordances affect one another reciprocally; they coevolve. In actively responding to an affordance, an organism creates new affordances. The affordances that biological form encounters constantly shift as form changes. The relation between an organism and its affordances is of the sort that Levins and Lewontin (1985) (following Engels) call 'dialectical':

These are properties of things we call dialectical: that one thing cannot exist without the other, that one acquires its properties from its relation to the other, that the properties of both evolve as a consequence of their interpenetration. (Levins and Lewontin 1985: 3)

A couple of examples might help to illustrate this reciprocal dependence and its importance for adaptive evolution.

The origin of hominoid tool use The advent of tool use in hominoids has been integral to their evolution, especially in late hominine lineages leading to *Homo sapiens*. The capacity of our hominoid ancestors to use tools has been intimately involved in human cognitive, linguistic and social evolution (Gibson 1994). But here there is a slight paradox. The capacity to benefit from the presence of tools in one's environment depends upon the capacity to use things in one's environment *as* tools. But if hominoid ancestors were incapable of tool use, how did the features of their environments advantage them *as tools*? Recent work in Evolutionary Developmental Biology suggests that it was changes in biological form instigated by other evolutionary pressures that initiated the capacity for tool use. The expansion of hominoid tool use may have been a contingent by-product of the evolution of obligate bipedalism (Rolian, Lieberman and Hallgrímsson 2010).

Tool use requires, at minimum, 'precision grip' (Marzke 1997). This is the capacity to oppose the thumb against one or more fingers.[14] Advanced tool use requires the ability to oppose the thumb against all of the fingertips ('higher order precision grip'). The ancestral prehominoid hand required changes in order for precision grip and opposition of the digits to be possible. The thumb

---

[14] For example the way you hold your door key between your thumb and the side of your first finger.

had to increase in strength and it had to be located more distally on the hand. In addition, the fingers had to be shortened (Rolian et al. 2010).

The developmental architecture of hands and feet are very similar. In fact hands and feet are serial homologues (Hallgrímsson, Willmore and Hall 2002). Campbell Rolian and colleagues point out that there is a great deal of correlation in the development of both (Rolian and Hallgrímsson 2009). Evolutionary changes in the development of one affect the development of the other. Rolian et al. (2010) demonstrate that the structural changes to the foot that facilitate endurance running are just those changes that in their homologous structures in the hand that are required for higher order precision grip: strengthening and distal extension of the big toe and the shortening of the lateral digits. They speculate that changes in the hand are a consequence of the evolutionary changes in foot structure. Given the developmental coupling of the hand and foot, changes in foot structure, as it were, drag the hands along.

[D]evelopmental constraints caused hominin fingers to evolve largely as a by-product of stronger selection pressures acting on the toes. Simply put, the shorter fingers and longer, more robust thumbs of humans likely evolved because of selective pressures on their respective homologues in the foot. (Rolian et al. 2010: 1564)

Nevertheless, these changes in hand structure conferred on hominoid ancestors new abilities to grasp implements and use them as tools. Changes in our ancestors' hands put tools in their environments. In other words, serendipitous changes in form dramatically altered the affordances of our ancestors' environments. These altered affordances, in turn, introduced new affordances, opportunities for adaptive evolutionary change.

The origin of metazoans   The changes in form that usher in new affordances do not have to be adaptive in any way, nor do they need to be underwritten by genetic changes. They can be induced by organisms' responses to a new physical environment. A vivid example can be seen in recent work on the origin of the Metazoa.

The morphological and developmental complexity of metazoans vastly exceeds that of any unicellular organism. Yet, the entire suite of basic metazoan structures, and a considerable amount of phyletic diversity, appears to have arisen rapidly in the Precambrian. How could such stunning diversity and complexity have arisen so precipitously? A fascinating picture is beginning to emerge (Newman and Bhat 2009). Kazmierczak and Kemp (2004) report evidence of a sudden rise in Ca^{++} concentrations in the Precambrian seas. This is important because increased Ca^{++} is known to promote cell-cell adhesion. The original coalescence of unicellular preanimals into vast assemblages of cells appears to have been the consequence of a change in the ionic constitution of the seas.

The nearest living relative of the metazoans appears to be the unicellular choanoflagellates (King 2004). Choanoflagellates possess a basic genetic tool-kit comprising (*inter alia*) genes that produce proteins that mediate cell-cell adhesions, genes that regulate growth and shape, and extracellular matrix proteins which – in metazoans at least – participate in cell sorting and tissue formation during development (King, Westbrook and Young 2008). The unicellular precursors of metazoans, then, carried genes that in the new context of multicellular assemblages played entirely new roles in metazoan function and morphogenesis.

[S]ome components of the protein machinery that mediates animal cell interactions may have originally played other roles in ancestral unicellular eukaryotes before being co-opted to function in signaling and adhesion. (King 2004: 319)

Newman reinforces the point.

What appears to have happened during the transition from choanozoan ancestors to metazoans is that pre-existing surface proteins began ... to mediate cell cell attachments. The consequent change in spatial scale created a context in which other pre-existing molecules were able to mobilize mesoscopic ... physical processes and effects that were irrelevant to objects of the size and composition of individual cells. (Newman 2011: 338)

The new arrival of massive aggregates of cells issues in a whole new range of capacities, challenges and opportunities. These proto-metazoans encountered 'mesoscopic' physical principles that had never previously affected the development or diversity of organic form. All these new affordances are a consequence of the forces acting on these new viscoelastic multicellular agglomerations.

The consequent change in spatial scale created a context in which other pre-existing molecules were able to mobilize mesoscopic (i.e. 'middle-scale') physical processes and effects that were irrelevant to objects of the size and composition of individual cells. (Newman 2011: 338)

The change to a multicellular context has numerous consequences (Newman and Bhat 2009): (i) while surface tension does not determine the shape of individuals cells, it does determine the shape of cell aggregates; (ii) cell aggregates containing surface polarising cells can acquire internal lumens; (iii) aggregates containing distinct populations of cells with different adhesive strengths will spontaneously sort out into layers; (iv) aggregates of cells that contain the same biochemical oscillators will spontaneously undergo synchronisation, in a way that allows the long-range co-ordination of cell activities; (v) cells that secrete diffusible molecules, when together in an aggregate, can act as sources of gradients that produce patterns of cells and tissues. Thanks to the newly encountered 'middle-scale' physical processes and effects, these aggregations of cells had the capacity to produce all the characteristic structures of the metazoans – lamina, vacuoles, blastocoels, tubes, metameres, differentiated tissues – spontaneously.

These new biological structures are immediate consequences of the physics of medium scale structures. They are not solutions to adaptive problems posed by an external environment. They just happened as a result of the physics of viscous, excitable materials.

[T]he forms of the earliest multicellular organisms . . . were more like certain materials of the nonliving world than are the forms of their modern, highly evolved counterparts, and that they were therefore almost certainly molded by their physical environment to a much greater extent than contemporary organisms . . . Stated simply, tissue forms emerged early and abruptly because they were inevitable – they were not acquired incrementally through cycles of random genetic change followed by selection. (Newman 2003: 221)

Nevertheless, they confer on biological form brand-new capacities, which in turn open up new vistas: threats to survival, opportunities for change, potential for new forms.

In each of our examples, hominoid tool use and the origin of metazoans, there are reciprocal cycles of changes in form with concomitant changes in affordances (without changes to the environment). But the changes in form that alter an organism's affordances are not merely passive or adventitious. The adaptive plasticity of development, identified in the preceding chapter as a manifestation of organismal purposiveness, is also a source of changing affordances. The plasticity of development alters form in an adaptively biased way in response to its affordances. As it does it also changes its affordances.

Adaptive evolution is not most perspicuously described as the process of form solving adaptive problems set by the organism's physical environment, but as form creating and then responding to an ever-changing system of affordances. The concept of an adaptation as a response to a set of affordances co-created by the organism and its setting makes salient the contribution of organismal form to adaptive evolution.

8.4   Situated adaptationism

I shall call this alternative to the adaptationist programme 'situated adaptationism' (Walsh 2012a). It has none of the defining features of orthodox adaptationism. Crucially, it does not posit a causal decoupling of organisms and the conditions to which they are adapted.[15] Consequently, it does not subscribe to explanatory

---

[15] At least insofar as the organism and the environment *are* causally decoupled, it is generally not of much explanatory significance. Organisms and their *experienced* environments are not so decoupled.

externalism. Because of the commingling of organism and environment, the process of generating a novel phenotype is *eo ipso* the process of generating a novel affordances. Adaptive responses to the affordances of an organism's niche not only alter *organisms* they alter their affordances as well.

8.4.1   Construction

If evolution is a response to affordances and yet affordances are partly constituted of the properties of organisms, this raises a problem for the very idea of form as *an* adaptation. The very concept of *an* adaptation, as Lewontin points out, seems to presuppose the decoupling of organism and environment.

To make the metaphor of adaptation work, environments or ecological niches must exist before the organisms that fill them. Otherwise environments couldn't *cause* organisms to fill those niches.

The history of life is then the history of coming into being of new forms that fit more closely into these pre-existing niches. (Lewontin 2001c: 63)

But this raises a dilemma. On one hand:

[S]o long as we persist in thinking of evolution as adaptation, we are trapped into an insistence on the autonomous existence of environments independent of living creatures. (p. 63)

On the other:

If ... we abandon the metaphor of adaptation, how can we explain what seems the patent 'fit' of organisms and their external worlds. (p. 63)

By way of pointing to a solution, Lewontin reminds us that

'[T]he environments of organisms are made by organisms themselves as a consequence of their own life activities. (p. 64)

Lewontin's message is that while the adaptedness of an organism (or one of its traits) is assessed as its suitedness to its conditions of existence, we must take into account the fact that those conditions of existence are very largely of the organism's making.

This is an old idea.[16] There are even traces of it in Darwin's version of Lyell's 'Economy of Nature' (Pearce 2009). Nevertheless, it has only recently begun to be accorded the importance it deserves (Odling-Smee, Laland and Feldman 2003; Turner 2000). The current buzzword for this relation between organisms and their environments is 'construction' (Odling-Smee, Laland, and Feldman 2003). Despite the attention, it isn't clear what it means to say

---

[16] It has certainly long been championed by Lewontin (1978, 2001a, 2001c), and has been reiterated many times (Griffiths and Gray 1994, Oyama 2000, Thompson 2007).

that organisms 'construct' their niches. It is possible to make out two distinct variants: the causal and the constitutive.

8.4.2   Churchill or Marx?

According to the causal construal, organisms cause changes to the features of the environment external to them. The activities of organisms effect some tangible alteration to the physically specifiable conditions around them. These modified physical conditions, in turn, redound upon the organisms. Beavers build dams; birds build nests; termites build mounds. These structures alter the organisms' physical environments (Turner 2000). Winston Churchill's quip that, 'We shape our buildings; thereafter they shape us' captures the essence of the 'causal' reading nicely.

It is well worthwhile paying attention to this 'Churchillian' niche construction. It is common, it certainly affects evolution and it is increasingly well understood (Odling-Smee, Laland and Feldman 2003). But it is also worth noting that Churchillian niche construction raises no particular difficulty for the adaptationist programme. On this construal, organisms and environments are still decoupled. But just as environments cause changes to form, organisms can alter the physical features of their environments. Nevertheless, when considering the implications of these alterations to the environment, there is nothing in the standard (Churchillian) approach to niche construction that precludes our treating these altered environments entirely independently of the organisms that altered them.

In fact, Churchillian niche construction is enthusiastically embraced by adaptationists. The very cases cited by those who promote the importance of niche construction for adaptive evolution – beaver dams, termite mounds, birds' nests – are endorsed by those who extol the enormous power of replicators. The 'long reach of the gene' ensures that genes build beaver dams, and termite mounds and birds' nests, just as surely as they build beavers, termites and birds.

A beaver dam, and the lake it creates, are true extended phenotypes insofar as they are adaptations for the benefit of replicators ... that statistically have a causal influence on their construction. What crucially matters ... is that *variations* in replicators have a causal link to *variations* in dams such that, over generations, replicators associated with good dams survive in the replicator pool at the expense of rival replicators associated with bad dams. (Dawkins 2004: 379; emphasis in original)

Sterelny (2005) presses the claim that the mere fact that organisms and environments exert reciprocal causal influences on one another shouldn't alter adaptationist thinking. He offers the following response on behalf of replicator-centred adaptationism to the suggestion that nest builders construct their environments.

From this perspective, nest evolution presents an evolutionary problem of the same kind as other complex adaptations.... However, if the evolutionary dynamics of extended phenotype adaptations are just like those of other complex adaptations, then the evolution of nests and burrows is not an instance of agent's [*sic*] changing their environments, it is simply an instance of adaptation to the environment. (Sterelny 2005: 29)

Sterelny concludes:

[T]he world without can be designed: it can be an extension of gene power. Genes act on their environment to improve their prospects of replication, and from the perspective of the gene, cells tissues, nests, burrows are just more environment. (Sterelny 2005: 30)

After all, if organisms are nothing other than structures built by replicators to alter their conditions of existence, nothing precludes extra-organismic features from playing the same role.

So, Churchillian niche construction is consistent with both the suborganicism and the explanatory externalism of the adaptationist programme (Godfrey-Smith 2001a). Niches may not be causally autonomous of the organisms in them, but they are still, on this view, *explanatorily autonomous*. The environment selects those replicator-built organisms that are most adept at living in it. This fact is quite independent of the question of how the environment got that way.

But, Lewontin's insistence that organisms construct their environments can be interpreted in a different way, one that is less obviously congenial to adaptationism. If causal niche construction has a Churchillian ring, of reciprocal causal construction, Lewontin's own formulation has distinct echoes of Marx's dictum that 'The animal is immediately one with its life activity. It does not distinguish itself from it ...' (1844).[17]

On this version of construction, organisms are commingled with their 'niches', not set apart from them. Organisms and their niches thus constitute a single interacting system in much the way that agent and environment do on the 'embedded/enactive' or 'situated' approach to mental phenomena advanced by Merleau-Ponty and Gibson.[18] The relation between organismal form and the conditions in which it evolves is dialectical. If the Churchillian conception of the organism/niche relation is that of reciprocal causation, the Marxian conception is that of reciprocal constitution.

There are many ways in which an organism can constitute its niche. It can effect a change to its environment's physical parameters, which changes in turn affect the organism itself, as in the Churchillian case. It can make a change

---

[17] Quoted in Ingold (1986).

[18] A number of accounts converge on this conception of the organism and its environment. These include (Griffiths and Gray 1994; Oyama 2000; Varela, Thompson and Rosch 1991).

in its own form, *without* affecting the environment, which in turn alters the affordances provided to the organism. The changes in the morphology of our hominoid ancestors' hand facilitated tool use, and the aggregation of single-celled proto-metazoans that facilitated the spontaneous formation of the array of typical metazoan macrostructures offer vivid examples. Alternatively, an organism can make itself differentially sensitive to its conditions (e.g., behavioural or physiological thermoregulation).

Constitutive niche construction underscores the reciprocity between organisms and their affordances that does not hold between organisms and their environments. What a feature of the environment affords an organism depends upon the organism's capacities, and the capacities of the organism in turn depend (in part) on the features of the environment. The constitutive notion of niche construction, offers us an opportunity for a constructive rethink of the Modern Synthesis conception of adaptive evolution.

Adaptive evolution is not the phenomenon that the adaptationist programme takes it to be. It is not the moulding of passive form by the exigencies of the external environment. This is precisely what the process of adaptation looks like if we omit the active participation of organisms. As a corrective, I suggest that we replace traditional Modern Synthesis adaptationism with an alternative that accords to organisms their rightful place in constituting and responding to the conditions under which form evolves. In what remains of this chapter, I shall explore a few implications of this 'situated adaptationism'.

8.5   Adaptation without adaptationism

In Chapter 6, we saw that orthodox Modern Synthesis evolutionary thinking not only dissociates suborganismal processes from extraorganismal (environmental) conditions, it puts them in competition for explanatory relevance. On that picture, the external environment causes adaptation (Lewens 2004). The processes occurring 'within' organisms – inheritance and development – are explanatorily relevant only insofar as biological form *isn't* adaptive. The supposition is that inheritance and development are essentially conservative; they 'constrain' form in various ways. Selection, on the other hand is 'creative' (Mayr 1976). Thus, we have two kinds of explanations; 'externalist' explanations of the adaptedness of form and 'internalist' explanations of the constraints on form. A recent articulation of this division of explanatory labour is found in Lewens (2009b). He argues that the commonality of form between related lineages is to be explained by the conservativeness of 'internal' development, while the differences in form within a lineage are to be explained by the external 'forces of selection'. The very idea that there are internalist and externalist explanations, rests on the traditional error of supposing that the influence of the environment on form and the effects of inheritance/development can be differentiated and apportioned.

8.5.1   Internalism and externalism

Many of the debates in current evolutionary biology and its philosophy take the form of a turf war, an attempt to carve out a larger territory for either the 'internalist' explanations that advert to the inner processes of inheritance and development, or the 'externalist' explanations that advert to the selecting influence of the environment. Ron Amundson points out that this is really just a warmed-over version of perhaps the oldest issue in comparative biology, the contrast between 'functional' and 'structural' approaches to explaining form.

Functionalism is the view that function is in some sense prior to form; objects have their form in virtue of the functions serve. . . . Structuralism is the view that form or structure is prior to function. Form, to a structuralist, is not to be explained in terms of function, but rather in terms of autonomous, formal/structural properties and processes. (Amundson 2001: 307)

Adaptationists are the inheritors of functionalism, a line that can be traced back to Spencer, Cuvier, Buffon. Those who accord the processes of development a prominent explanatory role are cast as the modern heirs of structuralism (Webster and Goodwin 1986), whose intellectual ancestry hails back to Geoffroy, Owen, and D'Arcy Thompson, Goldschmidt (Amundson 2005).

The dialectic of opposition has left an indelible mark on the history of evolutionary biology. Yet, from the perspective of situated adaptationism, the Panglossian dialectic of 'inner processes' and 'outer forces', of 'structure' opposed to 'function', of internalist explanation versus externalist explanation, looks quaint and misguided. For one thing, development and inheritance are not 'inner processes'. They are distributed throughout the organism/environment system, as Chapters 4 and 5 attest. Nor are the 'function' promoting influences on form strictly *external* to organisms. They too are spread throughout the organism/environment system.[19]

Nor is there, properly speaking, a division of explanatory labour between the processes that conserve form and the processes that promote its adaptive change. The activities of organisms – their capacity to exploit and alter their affordance landscapes – account for *both* the conservativeness of form and its capacity for adaptive change. Adaptive evolution is not the process by which the inner forces of organismal conservatism confront the outer forces of change. It is the process by which the adaptive plasticity of organisms conjointly moulds both the organism and its affordances, by affecting changes in organismal form, the environment, or both.

---

[19] Chapters 6 and 7.

Understanding the nature of the organism's commingled relation with its conditions of existence is the key to understanding the significance of organisms for adaptive evolution. But a proper understanding of commingling requires the abandonment of the dialectic of opposition: of structure versus function, of inner versus outer, of the forces of conservatism forces ranged against the forces of change.

8.5.2   Adaptation, no problem

If evolutionary adaptation is not the processes by which biological form is moulded to solve the pre-existing problems posed by the environment, then the concept of a biological adaptation bears no particular resemblance to the concept of design. A design *is* a solution to a pre-existing problem, a biological adaptation isn't (Griffiths and Gray 1994).

The Modern Synthesis conception of an adaptation, as a solution to a design problem is pervasive. It derives its intuitive appeal from the analogy between organisms and machines. Certainly, the power of Paley's argument from design resides in its appeal to machines; Paley (1809 [2006]) likened an organism to a watch. Adaptationist versions of the argument from design have substantially the same character. Organisms are cast as 'survival machines' built to do their replicators' bidding (Dawkins 1982).

But the machine analogy is misleading; organisms are not very much like machines. They are not the clocks that provided mechanism with its most potent guiding metaphor. They are not the watches that Paley used as his paradigm of design (Nicholson 2013, 2014; Talbot 2013). Organisms are Schrödingerian 'order-from-disorder' entities: self-building, self-nourishing, self-regulating, negentropic, homeostatic, goal-directed systems, in constant energetic commerce with their physical environments (Barandiaran, Di Paolo and Rohde 2009; Depew and Weber 1995; Varela, Maturana and Uribe 1974). Watches aren't. It is the very properties that set organisms *apart* from run-of-the-mill machines that confer on them 'a kind of *commingling* or *integralness* of [organism] and world – that is to undermine their very distinctness' (Haugeland 1998: 208). The machine analogy underestimates the significance to the process of adaptive evolution of the distinctiveness of organisms. In this it does more harm than good. It aids and abets the adaptationist programme's marginalisation of organisms. In drawing our attention away from organisms as adaptive agents, it obscures the causes of adaptive evolution from our view, and it promotes a false theory of evolutionary change. Adaptation is not design precisely because organisms are not machines in any run-of-the-mill sense. Thus, thinking in terms of the affordances and the rich sense of constitutive niche construction defended here requires us to reject the idea that organisms are designed machines. Rather, they must be seen as agents, making a place for themselves in the world.[20]

Conclusion

Adaptive evolution is the process in which populations change as a consequence of organisms responding to, and constructing, their affordances. An affordance, unlike an environment, is an emergent property of a purposive system embedded in its setting: an opportunity for, or an impediment to, the achievement of a goal. A system that responds to affordances is one that produces changes in its own state, or to its environment because, by and large, those changes are conducive to the attainment of the system's goals. Those changes, in turn, can be explained by citing their contribution to the fulfilment of those goals. The implication of the affordance concept is that certain features of evolution occur because organisms make them happen in pursuit of their goals.

To be sure there are advantages to this reconceptualisation, but it does not come cheap. Introducing organisms into evolution requires a significant deal of conceptual retooling. By comparison to the pared-down metaphysics of autonomous genes and environments, an ontology of affordances and co-constituting organism/environment systems seems positively profligate. The issue is that affordances imply purposes. Only a purposive entity can have affordances. That in turn introduces the spectre of teleology. We approach that issue in the next chapter.

---

[20] The machine-organism concept is deeply embedded in evolutionary thought (Nicholson 2013). Talbot (2013), for example, argues that machine-organism concept is responsible for the concept of the unit gene. Its abandonment has wide-reaching implications.

9 Natural purposes
Mechanism and teleology

In the late 1950s, David Rogers, a young medical researcher at Vanderbilt University, made a startling movie. At 28 seconds long, it is a marvel of microscopy for its time, and a wonder to behold at any time. It depicts a human neutrophil – a type of leukocyte – chasing a bacterium. Thomas Stossel (1999) posted the vignette online along with some running commentary: 'The neutrophil is 'chasing' *Staphylococcus aureus* microorganisms, added to the film. . . . As the neutrophil relentlessly pursues the microbe it ignores the red cells and platelets.'[1] Stossel's narration is heavily laden with purposive language, not the sort of talk to which biologists are readily disposed. But to witness this chase scene is to understand why it calls forth this kind of locution. The way the leukocyte tracks the bacterium, turns when it does, squeezes between the erythrocytes in pursuit, makes purposive talk wholly appropriate. The neutrophil really is pursuing the microbe; it really is ignoring the red blood cells. Sometimes the most apposite description of the biological world is teleological.

Teleology is a mode of explanation in which the existence or nature of an object or event is accounted for by citing the purpose it subserves. As we documented in Chapter 1, the worldview ushered in by the Scientific Revolution is inimical to teleology. That world is a realm of mechanisms and their effects: bodies in motion, the transmission of light, the flux of heat. As Molière attests, an enthusiasm for mechanism seems to go hand-in-hand with a disdain for purpose. The arch mechanist Frances Bacon deftly sums up the prevailing attitude. He scorns final causes as 'barren virgins dedicated to God'. Mechanism accords no place to teleology (or its shady, Scholastic sidekick, vitalism).

In its most austere and demanding forms, the mechanical philosophy insists on a disenchanted world explicable without remainder in terms of causal principles. Though mechanical philosophers differ from one another about which causal principles are basic ... they univocally reject explanations that appeal to vital forces and final causes. (Craver 2013: 133–134)

---

[1] The clip can be viewed online at: http://biochemweb.org/neutrophil.shtml.

Mechanistic philosophy was tailor-made for the dynamic new physics that inaugurated the Scientific Revolution. Despite the great optimism in which it was founded, it came with no guarantee that the new mechanistic methods should transpose smoothly over to the study of organisms. Indeed, vestiges of premechanistic science, like teleology and vitalism, loitered on the fringes of biology long after the rest of the world had been thoroughly mechanised. This is hardly surprising. The distinguishing features of organisms that we encountered in the preceding chapters – their self-building, self-organising, adaptive purposiveness – seem to require us to treat them as something more than mere congeries of mechanical parts.

Now if we look at the facts quite objectively, putting firmly out of our minds the mechanistic preconception, we find clearly exhibited in the living organism a mode of activity which is shared by no inorganic object or unit, and by no machine, namely, action directed towards end-states or goals . . . (Russell 1945: 3)[2]

The idea – obvious to many, rebarbative to just as many others – is that organisms are fundamentally purposive entities.

You cannot even think of an organism . . . without taking into account what variously and rather loosely is called adaptiveness, purposiveness, goal seeking and the like. (von Bertalanffy 1969: 45)

The relation between organisms and purposes is an intimate one. Gerd Sommerhoff points out that the capacities that distinguish organisms also serve as our criterion of purposiveness.

On the phenomenal level from which all science must proceed, life is nothing if not just this manifestation of apparent purposiveness and organic order in material systems. In the last analysis, the beast is not distinguishable from its dung save by the end-serving and integrating activities which unite it into an ordered, self-regulating, and single whole, and impart to the individual whole that unique independence from the vicissitudes of the environment and that unique power to hold its own by making internal adjustments, which all living organisms possess in some degree. (Sommerhoff 1950: 6)[3]

So, biology faces a quandary. It could hardly stake its place as a modern science in good standing while at the same time flouting its norms, in particular the stricture against explanation by purpose. And yet, teleology doesn't seem dispensable to biology. Even the committed mechanist, Kant, repines that there will never be a 'Newton of a blade of grass', someone who can explain biological form solely as a manifestation of mechanical law.[4]

---

[2] Russell makes this admission 'at the risk of being labelled a teleologist' (1945: 3).
[3] I thank Fermin Fulda for drawing my attention to this passage.
[4] As Marcel Quarfoord points out, for Kant 'teleology provides the objects of biological science' (Quarfoord 2006: 743).

Darwin's theory of descent with modification has often been heralded as the very thing that might bring biology into compliance with the mechanistic sciences. For his own part, Darwin took particular pride that his account of fit and diversity conformed to the model of *vera causa* explanations; these he took to be the *sine qua non* of the scientific method (Hodge 1987). He was not alone. As early as 1869 von Helmholtz (1971) praised Darwin for making biology a proper science by drawing it under the ambit of the mechanical philosophy.

Natural selection, so the thought goes, is a mechanism that explains the fit and diversity of organic form, and at the same time rids us of the murky metaphysics of purposes. Mayr holds that 'Darwin had solved Kant's great puzzle' (Mayr 1988: 58). Michael Ghiselin (1993) calls teleology a 'metaphysical delusion' for which the Darwin's theory of evolution is the most efficacious cure. David Hull, for his part, confidently proclaimed that 'From the point of view of contemporary biology, both vitalism and teleology are stone-cold dead' (Hull 1969: 249).

Modern Synthesis evolutionary thinking may have no truck with purposes and teleology, but we have also surveyed some reasons for exploring alternatives. The strategy of the foregoing chapters has been to argue that Modern Synthesis thinking is deficient, or incomplete, in various ways. It comes up short in precisely those places where our understanding of evolution could be augmented by taking into account the contributions that organisms as adaptive agents – that is to say, as purposive entities – make to the process of evolution. In the preceding chapter, for instance, I argued that adaptation, the process, is a consequence of the response of organisms to their affordances. Affordances imply purposes. In Chapter 7, I argued that the adaptive purposiveness of development is the process of cardinal significance in evolution.

That is all well and good, but natural purposes and teleology are generally repugnant to scientists and philosophers of a naturalistic bent. This nascent organism-centred evolutionary thinking will be nipped in the bud if it doesn't find a way to reconcile modern scientific methodology with teleology.

9.1   The case against teleology

On the mechanical worldview, to explain a natural phenomenon one adverts to the causal mechanisms that produce it. Mechanistic explanations are complete in the sense that citing causes is sufficient to account fully for the occurrence of an event. There is nothing left over for purpose to explain, once an event's mechanistic causes have been identified. They are exhaustive too, in the sense that every event has a complete mechanistic explanation. There are no gaps in the causal structure of the world for purposes to fill. Teleological explanations are thus not only a medieval throwback, steeped in a dubious metaphysics of ends and goals, worse still they are completely otiose. There is no point to purpose.

It is incumbent on any claim on behalf of teleology to show that it doesn't make any illicit metaphysical commitments, and that, the completeness of mechanism notwithstanding, teleology has some indispensable explanatory role to play. The case against teleology is compelling. Yet, addressing the charges one-by-one will help build an understanding of teleology and purpose, and of how the natural sciences can avail themselves of genuine, legitimate teleological explanations. A battery of standard arguments against teleology all point to the oddness of goals as explanantia.[5]

The argument from nonactuality: Means *precede* their ends (goals), but in teleology, ends *explain* their means. At the time of the occurrence of the means, the goal is an unactualised, nonexistent state of affairs. Teleology seems to require backward causation or some form of causation by nonactualia states of affairs. But nonactual states of affairs don't cause anything; they can't as they don't exist. Spinoza complains that 'this doctrine concerning the end turns Nature completely upside down. For what is really a cause it considers an effect, and conversely, what is an effect it considers a cause'.[6]

The argument from intentionality: Nonactual states of affairs cannot cause anything, but mental representations of them can. One way to avoid the commitment to nonactualia is to suppose that teleology requires intentionality. Our paradigm of successful teleological explanation, after all, is intentional. The occurrences of actions are explained by the intentions of agents, and in the same way, the features of artefacts are explained by the intentions of artisans. But typically organisms do not have intentional representative states.[7]

The idea that purpose presupposes intentionality figures prominently in historical discussions of teleology. Perhaps the earliest comprehensive treatment of teleology is to be found in the *Timaeaus*. There, Plato argues that the ordered regularity of the world must be a consequence of the intentions of the demiurge. The other basic causal relation in the world – the 'wandering cause' – brings only chaos (Johansen 2004). Aquinas likewise insists that the order in the natural world manifests purpose, and that requires intention. 'Now whatever lacks intelligence cannot move towards an end, unless it be directed by some being endowed with knowledge and intelligence; as the arrow is shot to its mark by the archer'.[8]

---

[5] The following discussion draws heavily from Walsh (2008).
[6] *Ethics*, Appendix 1 quoted in Curley (1994: 112).
[7] Increasingly, however, biologists are beginning to ascribe attribute intentional psychology, desires and rational choices even to unicellular organisms and the cells of multicellular organisms (see Shapiro 2007, 2011).
[8] Aquinas, *Summa Theologica* (2006), question 2, article 3.

It is certainly true that intentionality provides our paradigm of teleology. Even Kant (1790 [2000]) admits that intentionality is our only model for understanding purpose. That suggests that teleological talk is only ever (at best) analogical, a conviction that persists today.

End-directed thinking – teleological thinking – is appropriate in biology because, and only because, organisms seem as if they were manufactured, as if they had been created by an intelligence and put to work. (Ruse 2003: 268)

The argument from normativity: Teleological explanation appears to have normative import. Here again the paradigm of intentional explanation is illustrative (Taylor 1963). To explain an action as the consequence of an agent's intention is to demonstrate that the agent was rationally required (or at minimum rationally permitted) to commit the act, given her goals (Davidson 1963). Rational actions are those that are pursued because the goal is perceived desirable, 'under the guise of the good' (Boyle and Lavin 2010). Generalising the model, a teleological explanation must explain that, given the system's pursuit of e as a *goal*, and given that c is a means to the attainment of e, the system *ought* to have produced c (Stout 1996).

Marc Bedau (1991, 1998) argues that because of the normativity of teleological explanation, goals can only play their presumed explanatory role if goals themselves have intrinsically normative properties: c construed as the means toward the attainment of some goal e could only be something that the system *ought* to produce if e is a state that the system *ought* to attain, but e could not be a state that the system ought to attain unless e were intrinsically good. But natural facts are not intrinsically evaluable.

So, a naturalised teleology faces some robust challenges. Not only must it rebut these arguments, it must show that, the completeness of mechanism notwithstanding, there is some explanatory work for purposes to do.

9.2   Chance and purpose[9]

Accepting teleology incurs the cost of acceding to purposes in nature. Perhaps that price is too exorbitant. But before we count the costs of teleology, it is worth noting that foreswearing it hardly comes for free. In his landmark book *Chance and Necessity: An Essay on the Natural Philosophy of Modern Biology*, Jacques Monod faces squarely up to the consequences of banishing purpose from biology. Monod identifies 'a flagrant epistemological contradiction' gnawing at the heart of evolutionary biology. He calls it the 'paradox of invariance', and he cautions: 'in fact, the central problem of biology lies with this very contradiction' (Monod 1971: 12).

---

[9] This section is largely based on parts of Walsh (forthcoming).

The trouble, as Monod sees it, is that 'Living creatures are strange objects' (p. 17). They exhibit two extreme – and contradictory – kinds of organising principles: invariance and purpose. Invariance consists in the '... ability to reproduce and to transmit *ne variateur* the information corresponding to their own structure: A very rich body of information ... which ... is preserved intact from one generation to the next' (p. 12). But organisms are purposive too. Their purposiveness consists in the evident fact that they maintain their viability by responding and adapting to their conditions.

The problem is that austere scientific methodology sits uneasily with immoderate biological reality. On the one hand, 'The cornerstone of the scientific method is the postulate that science is objective.' Objectivity entails '... the systematic denial that "true" knowledge can be got at by interpreting phenomena in terms of ... "purpose" '. On the other hand, 'Objectivity nevertheless obliges us to recognize the teleonomic character of living organisms, to admit that in their structure they act projectively – realize and pursue a purpose' (p. 12).

In order to resolve the tension we could pursue either of two strategies. We might try to explain invariance – in particular the high-fidelity of inheritance, the robustness of biological form, and the stability of DNA – as a consequence of organismal purposiveness: in this way, '... invariance is safeguarded, ontogeny guided, and evolution oriented by an initial teleonomic principle, of which all these phenomena are the purported manifestations' (Monod 1971: 24). Alternatively, we could seek to explain the purposiveness of organisms by appeal to invariance. By this strategy, '... all properties of living beings rest on a fundamental mechanism of molecular invariance' (Monod 1971: 116). He opts for the second. This should sound familiar. Monod is here explicitly endorsing Schrödinger's 'order-from-order' account of living things.

Unlike Schrödinger, though, Monod maintains that our hand is forced by the intransigent demands of scientific methodology. The 'objectivity of science', with its commitment to mechanism, and it aversion to purpose, dictates that we explain the apparent purposiveness of organisms in terms of molecular invariance.

[T]he postulate of objectivity is consubstantial with science; it has guided the whole of its prodigious development for three centuries. There is no way to be rid of it, ... without departing from the domain of science itself. (Monod 1971: 12)

Grounding all the properties of life in molecular invariance raises a complication. Invariance is stasis, but evolution is change. Moreover, *adaptive* evolution is biased change. There must be a source of new variants, and there must be a process that biases evolutionary change. We cannot suppose that the new variants that arise are biased in favour of the goals or purposes of organisms, on pain of violating the 'objectivity of science'. So, the source of evolutionary novelties, Monod insists, must arise from unbiased – *chance* – alterations to the invariant molecular structure that underlies organisms.

The initial elementary events which open the way to evolution in these intensely conservative systems called living beings are microscopic, fortuitous, and utterly without relation to whatever may be their effects upon teleonomic functioning. (Monod 1971: 118)

Monod's argument is fascinating. In demonstrating that the methodological commitment to mechanism requires the Modern Synthesis to explain organismal purposes in terms of molecular invariance, he *also* shows that the very same commitment entails an ineliminable role for chance in evolution. The commitment to ineluctable chance in evolution is a methodological one. Monod is in no way reluctant to embrace the fundamentally chancy nature of evolution.

[Mutations] constitute the only possible source of modifications in the genetic text, itself the sole repository of the organism's hereditary structures, it necessarily follows that chance alone is at the source of every innovation, of all creation in the biosphere. Pure chance, absolutely free but blind, at the very root of the stupendous edifice of evolution . . . (Monod 1971: 112)

Monod traces the roots of his mechanism to the Pre-Socratic Atomist Democritus. There he finds the justification for biology's commitment to ineliminable chance. The book's very title pays conscious tribute. Monod credits Democritus with the claim that 'everything in the world is the fruit of chance and necessity'. In the introduction we briefly discussed the Atomist doctrine that the macro-level regularities of the world, including biological regularities, are the result of chance encounters of atoms moving randomly in the void. Those aggregations of atoms that are stable persist. The properties of stable aggregates are the necessary consequence of the interactions of their atomic parts. Given chance and necessity, the world has no need for purpose.

In fact, this trade-off between chance and purpose has a long and involved history. Just as Monod takes Democritus as his philosophical avatar, Aristotle takes him as his philosophical adversary. Aristotle's theory of explanation is in part an attempt to show that the Democritean conception of explanation is deficient. It lacks the resources adequately to explain important natural regularities. Its deficiency shows up precisely in the fact that it forfeits so much to chance.

9.3   Teleological naturalism

Aristotle introduces his powerful, elaborate, and thoroughly unmodern account of explanation in *Physics Book II*. His doctrine of the four *aitea* sets out a distinctive kind of pluralism about modes of explanation.[10] Aristotle's position is developed in explicit opposition to a then prevalent monistic approach to explanation, that of the Pre-Socratic Atomists. The Atomist treatment of explanation is, for all intents and purposes, like that of contemporary mechanists. To explain the activities of a complex system one simply adverts to the way that the interaction of its parts necessitates the effect to be explained (Hankinson 1998). Aristotle thinks that the Atomist account is impoverished, and that the most obvious symptom of its inadequacy is that it erroneously leaves certain perfectly explicable regular occurrences unexplained. It makes them look like chance events.

He illustrates the idea with a story of a man 'collecting subscriptions for a feast' (*Physics II.5*). In Aristotle's parable, our protagonist meets a debtor at the market and collects some money owed to him. Aristotle stipulates that this is a chance encounter: 'He actually went there for another purpose, and it was only incidentally that he got his money by going there' (*Physics II.5*).

This occurrence has a complete Atomistic/mechanistic explanation. The modern day version would cite physiological details like the depolarisation of neurons, and the contraction of muscle fibres, the movement of limbs, that propelled each toward the *rendezvous*. Aristotle's point is that this encounter gets the same Atomistic/mechanistic explanation *whether it is a chance event or it occurred for a purpose*. So, the resources of Atomistic/mechanistic explanation do not distinguish between certain kinds of regular occurrences and mere chance events. In particular, they do not recognise the difference between chance occurrences and those that occur because they fulfill a purpose. In an important sense, that's a good thing. Every occurrence must have a mechanical cause, whether it occurred for a purpose or not.

Nevertheless, there is all the difference 'in the world' between chance events and purposive events. And a good theory of explanation ought to be able to tell them apart. Crucially, purposive events and chance occurrences have different modal profiles. Purposive occurrences are robust across a range of alternate initial conditions and mechanisms; chance occurrences are not. For instance, if this were a purposive encounter, we would expect it to be somewhat *insensitive* to initial conditions, like location. Had the debtor been somewhere else in the market, or at the bath, or at the barber, then the encounter might well have occurred anyway, only in a different place, by a different set of mechanisms. Our protagonist would be expected to have done whatever was necessary to ensure that the event

---

[10] There is a considerable amount of debate about how best to translate '*aitea*'. The standard translation 'causes' is apt to confuse. I suggest that we think of *aitea* as explanantia, in the sense of there being four kinds of facts that can figure as the explanans in an explanation. Alternatively, there are four kinds of relations between explanans and explanandum: efficient, material, formal and final.

occurred. In other words, the mark of a purposive occurrence is that the means counterfactually depend upon the ends. In contrast, chance occurrences are highly sensitive to initial conditions; were this a *chance* encounter, then had the specific spatiotemporal circumstances been different, had the mechanisms been different, then in all likelihood, the event in question – the exchange of money – would not have occurred.

Then again, purposive occurrences are highly *sensitive* to goals in a way that chance occurrences are not. If this were a purposive occurrence, then had the agent's goals been different – say to collect money from someone else, to avoid his friend, to walk the dog in the country – the encounter probably would have not occurred at all. But, as Aristotle suggests, given that this is a chance encounter, if the creditors' purposes in being in that very place had been different – to buy soap, to place a bet – then holding constant the mechanistic causes, the encounter would still have occurred.

Because of their counterfactual dependence on both mechanisms and ends, events that happen because they fulfil a purpose can be explained in two different ways; one that identifies the occurrence as resulting from mechanical interactions, the other that identifies it as occurring because it conduces to the fulfilment of a goal (Walsh 2012b, 2013b). Ignoring purposes induces a selective blindness to a whole class of explainable regularities: those that are held in place by the counterfactual dependence of means on goals. This is not just an error of omission. It also runs the risk, Aristotle contends, of leading us into a much more serious mistake, that of misconstruing purposive occurrences as blind chance. In order to account for these events properly, we need teleological explanations as well as mechanistic ones. That in turn requires us to show that there are purposes (goals) in the world, and that they genuinely can explain their means.

9.4   Goals

Goal-directed processes bring about and maintain stable end-states. Those end-states are their goals. A goal is thus a state that a goal-directed process is directed toward. That may sound a little tautological, but at least it's true. Moreover, it forms the basis of a fairly simple approach to the naturalisation of goals. If we can offer a non-question-begging account of goal-directedness that does not presuppose the concept of a goal, it can be used to say what goals are.

The mechanisms generating goal-directed activity have been the subject of a considerable amount of empirical research. An initial wave of interest in goal-directedness originated in the cybernetics research of the 1940s to 1960s. Later research into complex adaptive, self-organising systems (Kauffman 1995) has supplemented this understanding. This has been expanded by research in

Autonomous Systems research (Barandiaran, Di Paolo, and Rohdel 2009; Di Paolo 2005).

The basic concept uniting these approaches is that of an adaptive, or autonomous system. Such a system is capable of attaining and maintaining robustly persistent states by the implementation of compensatory changes (Rosenbleuth, Wiener and Bigelow 1943; Sommerhoff 1950). Complex adaptive systems have the capacity to pursue a goal-state and sustain that state despite perturbations (Kitano 2004). Typically, an adaptive system is capable of implementing changes to its component processes in a way that corrects for the effects of perturbations that might tend to divert the system from its goal, or stable end-state.

A system's architecture underpins its goal-directed capacities. Such systems typically comprise arrangements of modules. We have seen the importance of modularity for organismal development (Chapter 7). For related reasons, it is significant to the understanding of goal-directed systems. Modules are clusters of causally integrated processes that are decoupled from other such modules. Modules exhibit a capacity to produce and maintain their integrated activities across a range of perturbations and influences; they are robust (Kitano 2004). Each module exerts regulatory influence, in the form of positive or negative feedback, over a small number of other modules. Complex, self-organising systems are usually arranged as hierarchies of modules. Kauffman outlines a set of general architectural properties shared by such systems; 'each part must impinge on rather few other parts' (Kauffman 1993: 67).

The consequence of this architectural structure is that complex adaptive systems are buffered; they are capable of maintaining their functional integrity across a range of perturbations. They also exhibit plasticity, the capacity to maintain stability in the face of perturbations by implementing novel stable adaptive responses. Robustness and the capacity to accommodate to novel circumstances in novel ways are the hallmarks of complex adaptive systems. Organisms, of course, are the very paradigms of complex, adaptive, self-organising systems. It appears that principal architectural feature that confers this property on organisms is the modularity of their development (Schlosser 2002).

This rich tradition shows us that goal-directedness is an unproblematic causal consequence of the architecture of an adaptive system. It is also an observable feature of a system's dynamics. It consists in the capacity of a system as a whole to enlist the causal capacities of its parts and direct them toward the attainment of a robustly stable end-point. That end-point is the system's goal. Being a goal is not a mysterious intrinsically normative property of a state of affairs. It is a complex relational property, the property of being a state that a goal-directed process tends to attain and maintain.

So, goals are natural and observable. Teleology's reliance upon goals, there-fore, should offer no impediment to naturalism. Ludwig von Bertalanffy presses the case:

[T]eleological behaviour directed toward a characteristic final state or goal is not something off limits for a natural science and an anthropomorphic misconception of processes which, in themselves, are undirected and accidental. Rather it is a form of behaviour which can be well defined in scientific terms and for which the necessary conditions and possible mechanisms can be indicated. (von Bertalanffy 1969: 46)[11]

This systems theoretic approach may tell us what it is for a system to have a goal. As such, it offers an account of the conditions under which teleological explanations apply. But, importantly, it does not tell us about the *content* of those explanations. Most attempts to naturalise teleology do not pay particular attention to the difference between the content of an explanation and the conditions under which it applies. The difference is crucial in the case of teleology (Walsh 2014c). The conditions required for a teleological explanation can be given in strictly causal terms. One just has to specify what causes the goal-directed behaviour. In contrast, the *content* of a teleological explanation cannot be specified in strictly causal terms. A teleological explanation does not mention the cause of an event's occurrence. It mentions the goal to which it contributes. In a teleological explanation, the goal state figures *as a goal* in the explanans. No causal explanation does that. Teleological explanations, prop-erly construed, are not a species of causal explanations.

The standard approach to naturalising teleological explanation is to effect a sort of translation schema between teleological explanation and some causal counterpart, that is to say, to cast teleological explanations as a cryptic form of causal explanation. These usually involve citing the historical efficacy of natural selection (construed as a mechanism) (Millikan 1989b; Neander 1991; Ruse 1971). Teleological explanations, on this view, are really just disguised etiological explanations. While this strategy serves to evade any commitment to goals as goals, it risks losing the potential windfall to be had by a proper account of how goals can explain their means. For those less metaphysically squeamish, there is another, more satisfactory way of naturalis-ing teleology.

9.5   Teleological explanation

It is incumbent on the advocate of natural teleology to show how *goals* figure in a nonmechanistic, normative mode of explanation. Two questions must be answered: (i) How is that we can explain the occurrence of an event by citing

---

[11] See Ayala (1970) for an articulation of the claim that teleology is an empirically observable feature of the biological world.

the end to which it is a means?, and (ii) in what sense could this explanation not be given by citing the mechanism that caused it? The most effective way to answer the first question is to demonstrate that goals explain their means in a way that is exactly analogous to the way that mechanisms explain their effects. They appeal to counterfactual invariance relations. To answer the second, we simply need to show that they appeal to *different* invariance relations than do mechanistic explanations.

In Chapter 1, we explored a contemporary variant of mechanism that has been vigorously promoted in recent years (Bechtel and Richardson 1993; Craver 2007; Craver and Darden 2013; Glennan 1996, 2002; Machamer, Darden and Craver 2000). On this view, a mechanism is a kind of cause.

Mechanisms are entities and activities organized such that they are productive of regular changes from start or set-up to finish or termination conditions. (Machamer, Darden and Craver 2000: 2)

A mechanistic explanation works by showing how the characteristic activities of the parts in question produce the properties of the effect to be explained.

[E]xplanation involves revealing the productive relation. It is the unwinding, bonding, and breaking that explain protein synthesis; it is the binding, bending, and opening that explain the activity of Na+ channels .... (Machamer, Darden and Craver 2000: 21–22)

The activities *produce* the effects. The relation that the producing activities – 'binding', 'bending', 'opening' – have to their effects is one of counterfactual dependence: effects counterfactually depend on their mechanisms. It is in virtue of this counterfactual relation that we can explain effects by appealing to the activities of the entities that cause them. Woodward calls this relation 'invariance'.[12]

[T]he sorts of counterfactuals that matter for purposes of causation and explanation are just such counterfactuals that describe how the value of one variable would change under interventions that change the value of another. Thus, as a rough approximation, a necessary and sufficient condition for X to cause Y or to figure in a causal explanation of Y is that the value of X would change under some intervention on X in some background circumstances .... (Woodward 2003: 15)

Being the mechanism of some occurrence is the change-involving invariance relation *par excellence*. It is possible, however, that there are other kinds of explanatory invariance relations.

---

[12] 'Invariance' used in this way is not to be confused with 'invariance' in Monod's sense. In Monod's usage, 'invariance' denotes unchanging robustness. In Woodward's it denotes a constant relation between two ranges of values for variables. The duplication of terms is unfortunate.

9.5.1   Purposive invariance

Remarkably, the same kind of relation holds between a goal and the means to its attainment. As the foregoing account of purposes suggests, a goal is a counterfactually robust kind of difference maker. For any system with goal e_1 that produces an event m_1, that is conducive to e_1 under actual conditions, under different conditions it would produce a different event m_2 such that m_2 would, in those conditions, conduce to e_1. Under similar circumstances, were the system to have had *a different* goal, e_2, then it would have produced a different event, m_3, conducive to e_2. The relation between an end (goal) and its means is thus an invariance relation. It is exactly the obverse of the relation that holds between a cause and its effects.

If invariance underwrites the explanatory relation between cause and effect, then it ought to underwrite an analogous relation between goals and means. Just as we can say that a cause explains its effect because were the cause to obtain then so would the effect and were it not to obtain the effect would not either, we can also say that a goal explains its means because were the system in question to have the goal then the means would obtain, and were it not to have the goal the means would not have occurred. In this sense, goals explain in more or less same way that mechanisms do (Walsh 2008, 2012b, 2013b).

Citing an invariance relation may be necessary for offering an explanation, but it is not sufficient. The reason is that explanation is 'description dependent'. A good explanation enhances our understanding of the explanandum. It cannot do that unless the invariance relation between explanans and explanandum is described in the right way. It may well be, for instance, that your favourite activity produces my favourite effect, but knowing that may leave us none the wiser about the event's occurrence. An effective mechanistic explanation must do more than merely identify a productive relation; it must, further, describe it in an appropriate way. In describing the productive relation between a mechanism and its effect, we use causal dispositional concepts the − like *pushing, pulling, attracting, binding*.[13] The interesting thing about dispositional concepts is that they wear their productive relations on their sleeves. If you know what *pushing, pulling, or attracting* is, then you know what happens to Y when X pushes, pulls, or attracts Y. The 'thick causal' description discloses the kind or nature of the invariance relation between X and Y: it's a pushing, or a binding. And in this way it explains the effect.

In a teleological explanation, a special kind of descriptive vocabulary is also used. We use locutions like 'in order to', 'for the purpose of', 'for the sake of', . . . to signify that the effect in question is a goal, *and* that the means whose occurrence we are explaining is *conducive* to the goal. So, just as descriptions

---

[13] These are what Nancy Cartwright (2004) calls 'thick causal concepts'.

in the mechanical mode identify causes as *producing* their effects, descriptions in the teleological mode identify means as *conducing* to their ends. And just as producing is a special kind of modal relation that holds between a mechanism and its effect, conducing is a special kind of modal relation that holds between an end and its means. We successfully explain an occurrence mechanistically when we demonstrate that is produced by some particular mechanism. By extension, we successfully explain an occurrence teleologically when we demonstrate that it conduces to some particular goal.

Conducing isn't just causing. A means conduces to its ends only if it would robustly and reliably bring about the end, *ceteris paribus*, across a range of counterfactual circumstances. If Y is a goal and X causes Y, it doesn't follow that X conduces to Y. If X causes Y only through the confluence of the most unpredictable, and convoluted happenstance – through a 'deviant causal chain' (Davidson 1980) – X does not conduce to Y.

'Producing' and 'conducing' descriptions of an event are different. They carry different information. The 'producing' description specifies that the earlier event is the mechanism of the later. It tells us 'how' the later event occurred. The 'conducing' description signifies that the later event is a goal. In doing so it specifies 'why' the former event occurred; it occurred because it is an effective means for realising the goal. The relation of conducing captures the normative dimension of the relation between a goal and its means. Given that Y is a goal, X is appropriate under the circumstances, just if X conduces to Y.

The upshot is that one and same event can have both a mechanistic and a teleological explanation. One explains *how* it happened; the other explains *why*. They are complete, complementary and noncompeting. Each is complete in the sense that it does not need to be improved or augmented by the other. As Democritus suggests, we do not make a mechanistic explanation any more complete (*qua* mechanistic explanation) by adding that the effect in question is also a goal. That fact is strictly incidental to a mechanistic explanation. Conversely, we do not improve a teleological explanation of *why* an event occurred by adding an account of *how* it occurred. Mechanistic and teleological explanations are complementary in the sense that each tells us something different about the event being explained. Each locates the event as part of a different, counterfactually robust regularity. So, while adding that the effect is a goal doesn't augment a mechanistic explanation, and adding mechanistic detail doesn't improve a teleological explanation, knowing both the mechanistic and teleological explanations of the occurrence of an event provides a more complete understanding of the event than either would alone. Finally, mechanistic and teleological explanations are noncompeting in the sense that one cannot replace the other without explanatory loss. Given their completeness and independence, knowing both the mechanistic and (where applicable) teleological identifies it as an instance of two distinct kinds of robust regularity: mechanistic and purposive.[14]

9.5.2   The antiteleological arguments

The foregoing, I take it, is really just an updated rendition of the Aristotelian account of teleological (final cause) explanation, cast in the language of counterfactuals and invariance. It is fairly evident that teleological explanation, conceived in this way escapes the battery of standard objections. It requires neither causation by nonactualia nor irreducibly normative facts, nor does it presuppose intentionality.

The argument from nonactuality: Teleological explanation requires no causation by nonactual states of affairs. Having a goal is an occurrent dispositional, empirically observable property of things. The system's goal is the state that it tends toward in this particularly robust way. The architecture of goal-directed systems reliably and predictably brings its goals about. Thus the architecture enables the system to establish a robust regularity between its occurrent state and the attainment of its goal. That is enough to secure the counterfactual dependence of means on ends that is the hallmark of teleological explanation. That is all that is required to legitimate the explanatory appeals to goals. Goals may be nonactual states of affairs, but it is certainly not dabbling in the occult to invoke them. The argument from nonactuality errs in assuming that for nonactual events to explain their means, they must cause them. In doing so, it simply conflates causing and explaining.

The argument from intentionality: It is true that the paradigm of successful teleological explanation appeals to the intentions of an agent. Intentions may well be representations of an agent's goals. They are also manifested as an agent's *goal-directed activity*. It is underdetermined, though, which of these aspects of intentions – the intentionality or the goal-directedness – does the explanatory work. The Aristotelian conception of teleology suggests that it is the goal-directedness that takes explanatory priority. The reason is that nonhuman organisms have goals or purposes, just as rational agents do, even if they do not have intentional states. As Aristotle tells us, we do not need to suppose there is intentionality in order to apply a teleological explanation:

It is absurd to suppose that purpose is not present because we do not observe the agent deliberating. (Aristotle [1996] Physics II. 8)

---

[14]   This of course appears to violate Kim's Explanatory Exclusion Principle (1994a, 1994b), which says no event can have more than one complete independent explanation. Walsh (2013b) offers an extended discussion of teleology and explanatory exclusion.

The argument from normativity: The argument from normativity incorporates two distinct claims, that in a genuinely teleological explanation: (a) a goal's *being a goal* must enter into the explanation and; (b) the evaluative status of the goal (i.e., its being *good*) explains why the system *ought* to act in a way that brings the goal about. The first claim is true, but the second is false. Goals are not the kind of states that *ought* to attain *simpliciter*. A goal-directed system tends to do whatever is necessary, under the circumstances, to bring about the goal. That's just what it is to be a goal-directed system. So, to say that system S did x in order to achieve y, is just to say that under the circumstances, x was conducive to y. It does not require that y is objectively good or valuable. Nevertheless, an important feature of the presumed normativity of teleological explanations is preserved. On any occasion of a goal-directed system undertaking an activity, it makes sense to ask whether that activity is appropriate. The activity is appropriate, on this view, only if it is conducive to the attainment of the system's goals. The upshot is that, contrary to the argument from normativity, teleology requires no irreducibly evaluative states of affairs.

Aristotelian teleology, then, is proof against the standard battery of antiteleological arguments (Johnson 2005). In fact, it appears that the antipathy to Aristotle's teleology that figured so prominently in the motivation of modern mechanism is based upon a misunderstanding (Shields 2014). For Aristotle, teleology is neither intentional, nor transcendent. Nor is it a kind of causation as such. It arises out of the fact that the activities of goal-directed entities issue in observable, projectible regularities in the natural world. These regularities can be used to ground predictions and explanations, in just the way that we use other robust regularities.

We have got this far. Teleology is an acceptable mode of explanation. Yet, for all its naturalistic credentials, it is not just a subspecies of mechanistic explanation. A mechanistic and teleological explanation can explain different facets of the same phenomenon: the how and the why (respectively). Hence, they are complementarity and noncompeting. Despite their differences, mechanisms and purposes explain in very much the same way, by citing robust counterfactual dependences. The counterfactual dependence of effects on causes underwrites causal/mechanistic explanations, and the counterfactual dependence of means on their ends licenses teleological explanations. That being so, the modern conviction that all explanation is mechanistic explanation stands in need of revision.

That is at best an argument to the effect that teleological explanations are permitted, but it does not demonstrate that they are in any way needed. None of the foregoing establishes that teleological explanation makes an ineliminable contribution to evolutionary biology. The argument from the preceding chapter, that adaptation is a response to affordances, and affordances imply purposes, is suggestive. But there are more direct, more compelling ways to establish the indispensability of biological teleology.

9.6   Teleology in biology

It is usually supposed that if teleological explanations have any brief at all in biology, they only apply in special limited circumstances (Ayala 1970). They illuminate the activities of various goal-directed subsystems – like the endocrine, immune, and thermoregulatory systems – that are specifically 'designed' to respond adaptively to uncertain circumstances. Perhaps they are also appropriate to the explanation of organismal behaviours. Moreover, insofar as teleological explanations are applicable in these cases, it is only because these explanations are ultimately grounded in 'the mechanism of natural selection'. Endocrine, immune, thermoregulatory systems (among others), are purposive because they have been made that way by natural selection. So, while teleology is a useful way to account for the workings of various purposive systems operating within organisms, it has no place in evolutionary explanation. Purposiveness may be an *effect of* evolution, but the purposive dynamics of organisms have no *effect on* evolution.

Increasingly, however, evolutionary biologists are questioning this exclusion of teleology from the standard account of evolution (Newman 2011; Shapiro 2011). If the argument of Chapter 8 is correct, we should expect that the purposiveness of organisms leaves a significant mark on the dynamics of evolution. I offer three quick examples the significance of organismal purposiveness for evolution: (i) the generation of evolutionary novelties, (ii) the fidelity of inheritance, and (iii) the function of genomes.

9.6.1   Plasticity and novelty

The importance of organismal robustness and plasticity have been stressed repeatedly in this work so far. Robustness and plasticity are two sides of the same coin. They consist in the capacity of an organism to attain or maintain a stable endstate through adaptive responses to perturbations. Robust, plastic organisms produce the responses they do precisely because, under the circumstances, those responses are *conducive* to an organism's survival. Robustness and plasticity are thus manifestations of organismal purposiveness. Robustness and plasticity are increasingly thought by many biologists to be the fundamental defining feature of living things: 'Phenotypic plasticity is a ubiquitous, and probably primal phenomenon of life' (Wagner 2011: 216). So, plasticity is purpose and it is basic but so far that entails nothing about how organismal purpose contributes to evolution.

In fact, however, we have already seen in preceding chapters that it does. To recap, phenotypic plasticity makes two general kinds of contributions to adaptive evolution. The first is in the production of novelties. Its second significant role lies in the orchestration of an organism's various developmental systems during adaptive evolution. The evolution of complex adaptations requires a significant degree of co-ordination. A change in, say, the strength of an appendicular muscle requires a concomitant change in the load-bearing capacity of its associated bone, an increase in vascularisation and accompanying changes to the nervous and integumentary systems. All these changes must be co-ordinated, and mutually supporting. Phenotypic accommodation permits all these systems to respond to a phenotypic (or environmental) novelty in ways that secure the viability of the organism as a whole (Pfennig et al. 2007).

In sum, then, the capacity of organisms to respond to changes and perturbations in ways that preserve their viability is required to *explain* the origin and maintenance of novel phenotypic characters in evolution. Evolution is adaptive, because organisms are adaptive, goal-directed systems. Novel phenotypes, and the accommodations they induce, occur when they do *precisely because* they contribute to the organism's goals of survival and reproduction. These are not chance occurrences. We need to invoke the capacity of organisms to pursue goals in order to explain the origin of adaptive novelties.

Allowing that organisms, cells and genomes are purposive systems exposes the fact that the source of evolutionary innovation is not random. Organisms possess the fundamental property that Wagner (2014) calls 'innovability'.[15] The origin of novelties is biased by the capacity of organisms to mount adaptive responses to their circumstances precisely because those responses are *conducive* to the achievement of their goals. The source of 'every innovation' is not random mutation, but the reactive, adaptive response of an organism's myriad systems to influences from genes, cells, tissues and environments. Mutations may be random; they really are indifferent to an organism's viability.[16] But organismal responses to them are not.[17]

9.6.2 *The high fidelity of inheritance*

One of the cornerstones of twentieth-century evolutionary biology is the conviction that the intergenerational stability of phenotype required for cumulative

---

[15] Wagner (2014) notes that given the high dimensionality of genotype space, the structure of gene regulatory networks makes random mutation a highly efficient way for them to innovate by searching the space of stable phenotypes.

[16] Francesca Merlin (2010) deftly defends the Modern Synthesis conviction that mutations are a matter of 'evolutionary chance'. I agree mutations are probably 'undirected'. It doesn't follow, however, that evolutionary novelties are random.

[17] Nor as we saw in Chapter 5 are mutations the only source of evolutionary novelties.

phenotypic evolution is a function of the unchanging nature of genes or replicators. On this traditional view, genes are by their very natures, highly stable, and conservative. They are particularly resistant to alteration by any processes downstream of replication and translation. As Jacques Monod avers:

[T]here exists no conceivable mechanism whereby any instruction or piece of information could be transferred to DNA ... Hence the entire system is totally, intensely conservative, locked into itself, utterly impervious to any 'hints' from the outside world. (Monod 1971: 110)

Yet, it is becoming evident that the high fidelity of phenotypic inheritance owes itself very largely to two goal-directed capacities of organisms. The first is the capacity to detect, respond to and repair DNA lesions that would otherwise be lethal. The second is the capacity of organisms' developmental systems reliably to produce inheritable phenotypes across an enormous range of genetic and environmental variations.

An organism's DNA is buffeted by stresses and insults that cause significant structural alterations. In humans these lesions occur at a rate of roughly 150,000 per cell per day (Ciccia and Elledge 2010). If left uncorrected, the consequences of this damaged DNA can be severe.

These lesions can block genome replication and transcription, and if they are not repaired or are repaired incorrectly, they lead to mutations or wider-scale genome aberrations that threaten cell or organism viability. (Jackson and Bartek 2009: 1071)

Cells have elaborate systems for detecting the various kinds of lesions and mounting appropriate correcting or mitigating responses. These responses vary from the disruption of mitosis, to splicing and repair of DNA structure, to apoptosis (Branzei and Foiani 2008), and these responses are highly sensitive to the kind and degree of potential damage. The DNA Damage Repair system (DDR) must identify and assess damage, and implement the correct cascade of responses. The DDR is thus like the immune system, endocrine or thermoregulatory systems; it is a highly sensitive goal-directed apparatus. Its goal is maintenance of the structural integrity of DNA, and *that* goal explains its specific activities. It manifests the hallmark characteristics of a purposive system: a broad, acutely sensitive, biased repertoire, an invariance of means on ends, a highly plastic capacity to respond in the appropriate way to even the most unpredictable perturbations, an ability to maintain a stable working endstate.

The DNA damage response (DDR) is a signal transduction pathway that senses DNA damage and replication stress and sets in motion a choreographed response to protect the cell and ameliorate the threat to the organism. (Ciccia and Elledge 2010: 180)

We saw Monod pronounce that the only 'possible' way to explain the purposiveness of organisms and the high fidelity of inheritance is to take the stable, 'invariant' nature of DNA as a primitive. But DNA is not invariant in the way Monod supposes. Instead, its stability is maintained by the activities of a purposive system, the DDR (Shapiro 2014).

The integrity of the genome that is evidently crucial to the high-fidelity of inheritance is not a primitive feature of the structure of DNA. Rather it is the result of the dedicated workings of a highly sensitive, adaptive goal-directed system. DDR implements the specific activities it does precisely because those activities are conducive to the system's goal of maintaining the structure of DNA. An organism's goals explain the constancy of DNA.

The stability of gene structure thus appears not as a starting point but as an endproduct – as the result of a highly orchestrated dynamic process requiring the participation of a large number of enzymes organized into complex metabolic networks that regulate and ensure both the stability of the DNA molecule and its fidelity in replications. (Keller 2000: 31)[18]

Organismal goal-directedness has further implications for the high-fidelity of inheritance. In the heyday of gene-centred evolutionary biology it was generally considered that genes severally code for discrete characters: 'gene for . . .' talk is predicated on this notion. Nowadays it is thought that genes operate in suites of interactive, adaptive gene regulatory networks. Gene regulatory networks manifest an adaptive robustness. Wagner (2011) and Pfennig et al. (2007) demonstrate the enormous capacity of gene networks to maintain their characteristic outputs – heritable phenotypes – across an astonishing range of environmental perturbations and mutations. Gene regulatory networks produce their characteristic output by actively compensating for genetic and environmental uncertainties. It appears, then, that the intergenerational constancy of phenotype that is essential for adaptive evolution is secured in some significant measure by the adaptive, goal-directed nature of gene regulatory systems.

9.6.3   Reactive genomes

As the conception of the gene as an individual unit of phenotypic control recedes, it is progressively being replaced by a conception of the genome as a highly integrated, goal-directed, corporate entity, the 'reactive genome'.

[T]he transition that concerns us has involved genomes rather than genes being treated as real, and systems of interacting macromolecules rather than sets of discrete particles becoming the assumed underlying objects of research. (Barnes and Dupré, 2008: 8)

The genome, conceived in this way, is not so much a repository of information for building an organism. It is an open system that exploits all the various resources available to the organism in development. These resources are

---

[18] We encountered some of these in Chapter 5.

genetic, epigenetic, cellular, extracellular and environmental. The production of phenotypes is not the exclusive or privileged province of any of these developmental resources. Rather, all such influences contribute to the production of an organism, under the reactive guidance of the genome as a whole. Genomes respond to and integrate all these cues.

At the very least, new perceptions of the genome require us to rework our understanding of the relation between genes, genomes and genetics . . .it has turned our understanding of the basic role of the genome on its head, transforming it from an executive suite of directorial instructions to an exquisitely sensitive and reactive system that enables cells to regulate gene expression in response to their immediate environment. (Keller [forthcoming]: 3)

Moreover, genomes themselves are edited, restructured and rewritten by the various cellular contexts in which they find themselves. We encountered an extreme example in Chapter 5. *Oxytrichia trifallax* rewrites 90 per cent of its somatic genome. But less extreme cases of the regulation of genome structure by the cell are commonplace (Shapiro 2011). The genome is no longer seen as a program, but instead as a read-write system, reacting to and being affected by its organismal setting (Shapiro 2011).

The reactive, adaptive dynamics of genomes is yet another manifestation of biological purposiveness (Walsh 2014c). We cannot understand the way that genes, cells, whole organisms or environments contribute to development unless we understand how genomes react to these influences. That is to say, an understanding of genome function must be predicated upon the recognition that genomes are goal-directed, purposive systems.

These are but a sample of the many ways that the purposes of organisms, and suborganismal systems contribute to the process of evolution. Many more could be cited. No doubt there are many more yet to be discovered. My objective here is to raise awareness of the pervasive role of purpose in maintaining organismal viability, and in creating the conditions for biological evolution (Walsh 2006). The robustness of inheritance, the plasticity of development, the origin of novelties, the stability of the genome, the adaptedness of evolution: these are not matters of chance. Stuart Kauffman was an early exponent of the view that orthodox Modern Synthesis thinking has overplayed the hand of chance: '. . . the merging sciences of complexity suggest that order is not all accidental, that veins of spontaneous order lie at hand' (Kauffman 1995: 8). The spontaneous order of the biological world is held in place by the purposiveness of organisms, and it is crucial to the process of adaptive evolution.

Conclusion

Modern Synthesis evolution is 'chance caught on the wing' in Jacques Monod's resonant phrase. The Modern Synthesis commitment to chance falls directly out of its endorsement of mechanism as its guiding methodology, and its concomitant rejection of purpose. To be sure, a purpose-free mechanistic worldview has the benefit of parsimony. But, it isn't entirely clear that this bargain-basement metaphysics is up to the demands of explaining biological evolution. In particular, there are robust invariance relations that are the consequence of biological purposes. The mechanist's etiolated conception of nature leads us to overlook these. As E.S. Russell insists: '. . . such an approach leaves out of account all that is distinctive of life, the directiveness, orderliness and creativeness of organic activities' (Russell 1945: 3).

These purposes can figure in genuine scientific explanations. Teleological explanations account for the occurrence of an event by demonstrating that it conduces to the attainment of a goal. Goals can explain their means because goal-directed systems are endowed with the capacity to bring about those states of affairs (means) that are conducive to the attainment of their ends. Where goals are involved, some events are susceptible to both mechanistic and teleological explanations. While the teleological explanation cites the goal to which the occurrence conduces, the mechanistic explanation cites the mechanism that *produces* it. The teleological explanations are not reducible to, or eliminable by, their mechanistic counterparts.

If our scientific methodology fails to countenance purpose, then it renders us blind to a perfectly real, evolutionarily important class of empirical regularities. The consequence is that what ought to be explicable, is dismissed as unexplainable, ineluctable chance. This is a mistake. It is a methodological artefact. Evolution merely *appears* chancy if we disregard the role of purpose. Aristotle's approach to scientific explanation offers biology an escape from methodological straitjacket imposed on it by the strict adherence to mechanism.

10 Object and agent
Enacting evolution

Physarum polycephalum is garnering quite a reputation. It is an unprepossessing cellular slime-mould that forms a plasmodium. A plasmodium is a contiguous branching structure in which protoplasm flows freely throughout, transporting cellular materials and nutrients. *P. polycephalum* exhibits the sort of behaviours that have researchers exalting their 'intelligence'. They are capable of anticipating regular events (Tero et al. 2008). They are able to 'compute' the most efficient route through an arrangement for nutrient sources (Teru et al. 2010). They can 'choose' the optimally balanced diet (Bonner 2010). They are not the only organisms to which these sophisticated capacities are regularly attributed. Even much simpler bacteria are said to manifest behavioural abilities that tempt researchers to think of them as 'sentient beings' (Shapiro 2007). The most rudimentary organisms constantly respond to their circumstances in novel and adaptive ways. So supple are their interactions that it is wholly inadequate to describe these organisms as mere machines. Perhaps these imputations of full-blown cognition are excessive. Bacteria may not entertain propositional attitudes, there may be no ratiocination as such in slime moulds.[1] Nevertheless, even if organisms are not cognitive agents, it must be recognised that they are purposive systems, agents of a sort.[2]

I have been stressing that evolution is an ecological phenomenon. It happens to a population (or a lineage) as a consequence of individual organisms' purposive engagement with their conditions of existence in the struggle for life. That, at least, is what Darwin taught us. The last few chapters have attempted to fill out this notion of 'struggle' and to draw out some of its implications. The 'struggle' consists in an organism's pursuit of its particular way of life. The activities that make up this pursuit are not at all like the

---

[1] Increasingly, there is interest expressed in dropping the 'lower bound' of cognition, in order to include the kind of sensitive, adaptive agency exhibited by most unicellular organisms (Van Duin, Keijzer and Franken 2006). At the very least, it is becoming recognised that full-blown cognition is contiguous with the kinds of capacities that are constitutive of living things (Lyon 2005). See Fulda (ms) for an enlightening discussion.

[2] Recall Aristotle's claim that it 'is absurd to suppose that purpose is not present because we do not observe the agent deliberating' (*Physics* II.8).

activities of ordinary inanimate objects. Most significantly, as I argued in the preceding chapter, they are purposive. In Chapter 8, I motivated this organismal perspective by arguing that the unit of greatest theoretical significance for evolution is not the gene, or for that matter even the organism *per se*. It is the organism situated in a system of affordances. Affordances are emergent entities; they are properties of a system, in this case, a system comprising an organism and its conditions of life. Affordances are constituted in large measure by the ways that organisms can exploit or ameliorate these conditions. Purposiveness and affordances are related. Only purposive entities have affordances, and purposive activity is the response to, and the simultaneous creation of, affordances.

This is tantamount to saying that organisms are *agents* of evolutionary change. Therein lies a significant departure from the place accorded to organisms in orthodox Modern Synthesis thinking. In the Modern Synthesis, organisms are treated as *objects* of evolution. The objective of Modern Synthesis evolutionary biology is to explain organismal form as the result of evolutionary processes – selection, drift, mutation and migration – impinging on populations of genes. Very little attention is paid to the way that organisms participate in – indeed *enact* – the process of evolution. By 'enacting evolution' I mean that through their activities organisms create evolutionary change, and also the conditions to which evolutionary change is a response.[3] As agents, they 'actively participate in their own evolution' (Ingold 1986: 187). If we are to assimilate organisms as agents of evolutionary change into evolutionary theory, some revisions need to be made. The anthropologist Tim Ingold makes effectively the same point.

In place of the kind of 'population thinking' ... that is the hallmark of Darwinian biology it is necessary to substitute a kind of 'relationships thinking', which locates the organism ... as a creative agent within a total field of relations whose transformations describe a process of evolution. (1989: 208)[4]

Much has been said in the foregoing chapters – particularly Chapters 4 through 8 – by way of motivating this change of perspective. It receives its impetus partly from a perceived inability of orthodox Modern Synthesis thinking to accommodate the significant features of developmental plasticity and robustness into our understanding of evolutionary processes (Chapters 6 and 9). It is further motivated by a dissatisfaction with the way the Modern Synthesis represents the component processes of evolution, as fractionated and

---

[3] The notion of enaction is borrowed from Varela, Thompson and Bosch (1991).

[4] The context here is Ingold's attempt to integrate anthropology and evolutionary biology. He is concerned with the way that humans, as agents, structure the conditions of their own evolution. Ingold and I appear to differ in that I hold that incorporating organismal agency into evolution is wholly consonant with Darwinism (bit incompatible with the Modern Synthesis).

quasi-autonomous (Chapter 3). That theory, as discussed in Chapter 4, insists on a restricted, regimented conception of inheritance, in which inheritance consists exclusively in the transmission of replicators. That supposition, in turn, looks to be inadequate to encompass the panoply of ways that the *pattern* of inheritance – transgenerational resemblances and differences – can be realised. The Modern Synthesis further underestimates the range, kinds and sources of evolutionary novelties (Chapter 5). It also engenders a misleading interpretation of natural selection (Chapters 2 and 6), and adaptation (Chapters 7 and 8). Most especially, what John Maynard Smith took to be its cardinal virtue, that the Modern Synthesis theory made it '. . . possible to understand genetics, and hence evolution, without understanding development' (Maynard Smith 1982: 6), turns out to be its most conspicuous failing. Even if the empirical deficiencies are only now beginning to show up, I suggest that what has been missing all along from the Modern Synthesis is the perspective of organisms as agents of evolution.[5]

I have said little so far about the implications of adopting this agential perspective. I think they are significant. A theory that places organisms as the agents of evolutionary change is a very different kind of theory from the Modern Synthesis. It is not entirely clear that the Modern Synthesis could take on the perspective of the organism as agent without abjuring its most deeply held convictions.

10.1   Natural agents

The notion of a natural agent follows directly from the idea of a natural purpose explored in the preceding chapter. Agency, like purposiveness, is an observable property of a system's gross behaviour.[6] It consists in a capacity of the system to pursue goals, to respond to the conditions of its environment and its internal constitution in ways that promote the attainment, and maintenance of its goal states. Agency is observable in the sense that what we see when we observe an agent is its dynamics, the way that the agent negotiates its situation. We observe a range of robustly regular responses to its conditions. Understanding a systems goal, for example the goal of the leukocyte in Rogers's film, allows us reliably to predict its behaviour.

In observing agency, we are witnessing an 'ecological' phenomenon. By an 'ecological' phenomenon, I mean that agency consists in the capacity of a system to cope with its setting, to attain its goal by responding to its affordances *as* affordances.

---

[5] Godfrey-Smith (2009) calls evolutionary theory an 'agential' theory. We mean different things.

[6] I offer here only the most cursory sketch of organisms as natural agents. I understand well that naturalising agency is a challenging project. (See Barandiaran, Di Paolo and Rohde 2009; Di Paolo 2005; Ruiz-Mirazo and Moreno 2012; Ruiz-Mirazo et al. 2000).

There are the three key concepts required for giving an ecological account of agency: *goals, affordances* and *repertoire*. They form an interdefineable cluster. An affordance, as we have seen, is an opportunity for, or an impediment to, the attainment of a goal. Only a goal-directed system can experience its conditions *as affordances*. Moreover, a system can only experience its conditions as affordances if it has a repertoire. By repertoire I mean the set of possible responses that a system can enlist in pursuit of its goal (in response to its conditions). For this repertoire to constitute a response to *affordances* (as such), it must be biased. That is to say that the system must have the capacity to exploit its behavioural repertoire in response to its conditions in ways that are by and large conducive to the attainment (or maintenance) of its goal. For its part the goal of the system is that state that it tends robustly to attain, or maintain, by marshaling its behavioural repertoire, in response to its affordances. So, the concept of agency, as I am using it, is located within a triad of concepts: *goal, affordance, repertoire*.

Repertoires come by degrees. Some agents have much richer repertoires than others. There are only so many responses that a bacterium can make – it can respond adaptively to nutrient gradients, toxins, and in some cases magnetic fields (Dretske 1986). This is meagre compared to the repertoire of *P. polycephalum*, and miniscule compared to that of most metazoans. Systems with rich repertoires are capable of responding to wider ranges of affordances. These systems, in turn, are generally capable of pursuing a wider range of goals. Ecological agency, then, is not an all-or-nothing thing. It comes by degrees. There is a continuum from the most basic agents, those capable of pursuing only a narrow range of goals, through those possessing an enhanced repertoires of responses, like those of the cellular slime moulds and neutrophils, through an ever-increasing range of complexity through to full-blown cognitive agents like ourselves.[7]

I maintain that natural agency is metaphysically unproblematic, certainly no more problematic than natural purpose. It is a gross behavioural property of a system, an 'ecological' phenomenon comprising purposive entities situated in their affordances. However, admitting agents into our ontology raises methodological challenges. We must find a way to treat agents *as agents* within evolutionary theory.

10.2   Object theories and agent theories

Consider the differences between two kinds of theories. I'll call them 'object theories' and 'agent theories'.[8] Object theories are familiar to us, and agent

---

[7] These issues are discussed in depth by Fermín Fulda (ms). I am grateful to Fermín for help here.
[8] The distinction is anticipated, to some degree, in Lewontin's (1985) discussion of 'variational' and 'transformational' theories. Lewontin identifies Darwin's theory as one in which organisms are the 'objects' of evolutionary forces.

theories less so. In an object theory, the domain of interest is a set of objects. The goal of the theory is to describe and explain the dynamics of these objects. So, we set out a space of possible alternatives for those objects – a state space – and we look for principles that might account for various possible trajectories through the state space. The objects in the domain are subject to forces, laws and initial conditions. The physicist Lee Smolin (2013) has dubbed this conception of scientific theories, the 'Newtonian paradigm'.[9] In the Newtonian paradigm, we describe the dynamics of a system by answering two simple questions: '(i) what are the possible configurations of the system? and (ii) What are the forces that the system is subject to in each configuration?' (Smolin 2013: 44). Crucially, in the 'Newtonian paradigm' the forces, laws and initial conditions are extraneous to the objects, and exist independently of them.

Object theories are characterised by what we might call 'transcendence' and 'explanatory asymmetry'. By 'transcendence' I mean that the principles that govern the dynamics of the objects in the theory's domain are not part of the domain itself. They do not evolve as the system under study does. The laws of nature, and the space of possibilities through which the objects move remain constant as the objects change. In this way, we can explain the change state of the system under study by appeal to the unchanging laws. In Newton's account of the orbits of the planets, for example, the planets move, but the laws of gravitation and the structure of space stay the same. Because of this transcendence of the principles over the objects, there is an explanatory asymmetry. The principles – e.g., laws of nature, initial conditions, the space of possible configurations – explain the changes to the objects in the domain, but the objects do not explain the principles. We cannot look to the motions of the planets to explain why the laws of gravitation are as they are. There are strong echoes here of Lewontin's (2001c) claim that the Modern Synthesis conception of adaptation requires us to see the environment as wholly extraneous to, independent of, and 'acting on' biological form. Modern Synthesis explanations are, consequently, asymmetrical.

Whereas object theories are characterised by transcendence and asymmetry, agent theories are characterised by what I shall call 'immanence' and 'explanatory reciprocity'. In an agent theory the entities in the domain include *both* agents and the principles we use to explain their dynamics.[10] The agents' activities are generated endogenously; agents cause their own changes in state in response to the conditions they encounter. These conditions, in turn, are largely of the agent's making. So, as agents implement their responses to their conditions, they not only alter their own state, they also change the conditions to which their activities are a response.[11] More changes ensue.

---

[9] Alternatively, in the case of physics, 'physics in a box'.   [10] That is the 'immanence' part.
[11] That is the 'reciprocity' part.

There is thus a dialectical relation between the activities of the entities in the domain, and the principles we call upon to explain them. The activities of the agent can be explained as a response to its conditions, and reciprocally, the change in conditions can be explained as a consequence of the activities of the agent. The objective of an agent theory is to account for the interplay between the entities whose dynamics we want to explain and the principles we use to explain them.

The very idea of an agent theory may sound a little mysterious. Most of our successful theories are of the object type. And in the absence of a good, independent, example of an agent theory, it is hard to make it much more accessible. But an interesting proposal has recently come to light.

Lee Smolin (2013) has argued that a complete theory of the physical universe requires a significant departure from the orthodox approach that dominates current physics: 'the remedy must be radical, not just the invention of a new theory but . . . a new type of theory' (Smolin 2013: 250). The new type of theory he envisages is one in which the laws of nature and the principles that explain the dynamics of the universe evolve as the universe does, in such a way that each explains the other.

Smolin's proposal is that a theory of the universe must break free of the Newtonian paradigm, for a fairly simple reason. In an object theory, such as we find in the Newtonian paradigm, the laws and initial conditions are there to explain, but not to be explained. They are merely 'input': *givens*. The best that such a theory can do is tell us that *given* the laws and initial conditions, the universe should evolve in such and such a way. But, Smolin argues, there is an infinite array of possible laws and initial conditions. So, a theory of the universe in the Newtonian paradigm could not answer the questions 'why these initial conditions?', 'why these laws?'. Smolin argues that this is unsatisfactory as a theory of the universe because, these questions, presumably, have answers. But the answers do not fall within the ambit of the theory. The answers must look beyond the theory of the universe. Yet, as Smolin insists, if our objective is a complete theory, then '[n]othing outside the universe should be required to explain anything inside the universe' (Smolin 2013: 121–122). So, the laws and initial conditions must be within the domain of the theory of the universe. Smolin proposes that the laws of nature, and the conditions to which the universe responds as it grows and complexifies will themselves evolve, as a consequence of changes in the structure of the universe. Smolin is calling for an agent theory of the universe.[12]

Modern Synthesis evolutionary theory is an object theory. It conforms to the Newtonian Paradigm.[13] The domain of objects comprises genes. The space of

---

[12] I thank Lee Smolin for a very helpful conversation on this issue.
[13] Indeed it seems that its advocates have spent much of its history trying to 'Newtonianise' it (see Sober 1984; Stephens 2004).

alternatives is represented as a fixed landscape of genotype spaces and their fitnesses. Populations are propelled through this space by extraneous forces of selection and drift.

The conviction that selection explanations are externalist and asymmetrical (Chapter 8) is an expression of this idea that the principles that explain evolutionary change are extraneous to the objects of evolutionary theory. Like any theory in the Newtonian paradigm, there is much about the process of evolution that is left unexplained by the theoretical apparatus. The range of variants available to selection is determined by initial conditions. The structure of the fitness landscape is also part of the initial conditions. One conspicuous lacuna is the origin of evolutionary novelties. Novel variants, mutations, are introduced randomly, as mere 'inputs' – *givens* – that fall beyond the purview of evolutionary theory.

There is nothing about the objects of the domain – genes – that answers Smolin's two questions: (i) what are the possible configurations, and (ii) what are the forces. The possible configurations comprise the set of possible genes and their combinations in organisms. The distribution of genes in a population do not determine this space of possibilities. Likewise, the forces that bring about the various configurations, selection, drift and mutation happen to genes, but their actions are dependent upon factors extraneous to the genes themselves.

If orthodox Modern Synthesis thinking is an object theory, then Situated Darwinism is an agent theory. It takes evolution to be the consequence of the purposive activities of organisms. It certainly exhibits the immanence and reciprocity that is diagnostic of agent theories. The capacities of organisms to succeed in the struggle for existence are both the phenomenon to be explained, *and* the principles that explain them. The conditions that propel a population through its evolutionary changes – the affordances – are not extrinsic to organisms. They are largely of the organisms' making. As organisms respond, they alter those conditions, which in turn causes organisms to respond. Form and its conditions evolve reciprocally. The conditions explain the evolution of form, and changes in form explain the evolution of the conditions. Similarly, the range of variants and the space of possibilities available to a population are also largely of organisms' making. At any stage the range of phenotypes available to a population depends upon the phenotypic repertoires of the organisms in it.

10.3   The disappearing agent

The fact that organisms are by their natures agents, and that the Modern Synthesis is an object theory, helps to explain why organisms have disappeared from evolutionary theorising. As we discussed in Chapter 1, the whole edifice of modern science, from Descartes onward, has been dedicated to expunging agency (at least from the sciences of noncognitive domains). But if agency is a real, natural phenomenon, and our scientific theories cannot countenance it, then our understanding of the world is destined to be impoverished. One possible strategy in response might be to try to enrich our orthodox Modern Synthesis theory by grafting onto it the fact that organisms are agents. I don't hold out much hope for a mere addendum to the existing theory. I'm inclined to think that not only don't object theories acknowledge agency, they *can't*.

This inability is clearly on display in an area of philosophical discourse that offers an illustrative analogy – the treatments of rational action. Action theory is torn between two conceptions of humans, as objects in the natural world, subject to exogenous causal influences, and as agents, capable of initiating actions guided by reasons. The parallel between action theory and agent-centred evolutionary thinking is salutary.

10.3.1   Action theory

In action theory, as in biology, agency seems at once to be real, and beyond the purview of our best naturalistic theories.[14] Jennifer Hornsby (1997) and David Velleman (1992) (amongst others) have expressed the concern that standard approaches to explaining human action have left us with the problem of the 'missing agent'. The 'missing' or 'disappearing' agent is a consequence of a set of methodological commitments that are commonly enlisted to explain action. These by now are familiar to us: they arise from the precepts of Cartesian mechanism. The Cartesian mechanist version of action theory holds that an agent's thoughts – beliefs and desires – explain her actions only if they *cause* her actions (Davidson 1963). In much of contemporary action theory, this is interpreted as implying that thoughts are mental entities realised as internal physiological mechanisms, and that these mechanisms combine with other internal mechanisms to effect actions. They do so by dint of their intrinsic causal properties (Fodor 1987).

Actions are outputs of something like an internal process of computation. They result from the mechanical interactions of subagential, internal states of the agent.[15] The purposes of agents, their own agential dynamics, do not appear in the explanations of actions. On the Cartesian model of thought and action, agents are middlemen, the interface between the causal activities of their subagential physiological states and the demands of the environment with which the agent must deal.

---

[14] Again, I thank Fermìn Fulda for helpful discussions on these issues.
[15] In Chapter 8 we noted that this way of thinking is strongly analogous to the view that the properties of organisms are due to the interaction of suborganismal mechanisms.

The worry expressed by Hornsby and Velleman and others is that this picture risks losing sight of the commonsense intuition that actions don't just *happen to* agents; agents *perform* actions.[16] Cartesian mechanism misses the whole point of a theory of action. An action is something that is produced *by* an agent, *for* a reason. The objective of a theory of action, properly construed, is to explain the doings of agents by showing them to be reasonable, or rationally justified, given the agent's purposes or intentions. A successful action explanation demonstrates that, given the agent's goals, under the circumstances, her actions were appropriate (or otherwise). In this context an action is appropriate just if it is conducive to the fulfilment of the agent's goals. Taking actions to be mere causal consequences of the machinations of suborganismal states induces us to miss entirely the very thing that makes actions *actions*. Rational agents (like organisms) are purposive entities and actions are the manifestations of their pursuit of their goals, or intentions.

The problem here is that the Cartesian mechanist approach is an object theory. It takes agents and actions as objects, to be explained by extraneous forces or causes acting on (or producing) them. It attempts to explain actions not as the autochthonous products of agency, but as the effects of extrinsic causes: external environments on the one hand and internal computation and representation, on the other.

There is a live alternative to psychological Cartesianism. A general class of approaches to mind posits agency as primitive (Taylor 2005; Thompson 2007). For example, Merleau-Ponty's account of the explanation of behaviour begins with the agent as an active, problem solving, goal pursuing organism (Matthews 2002). The agent responds to her conditions as meaningful – they offer impediments to or opportunities for the fulfilment of her goals. The sense in which an agent's conditions are 'meaningful' should be familiar to us. An agent's goals and capacities imbue her conditions with significance. She responds to them as threats to, or opportunities for, the fulfilment of her goals. An action, then, is a response initiated by the agent, to a set of affordances, in pursuit of her goals. The affordances are to a significant degree of the agent's making, and they evolve in concert with her actions and goals.

This is an agent theory of action. It casts actions as events generated by agents, that occur as a consequence of agents' pursuit of their own purposes. These purposes, moreover explain and justify the actions. This suggests that the explanation of action, *as* action, requires an agent theory. As for action, so too for adaptive evolution. Adaptive evolution is a phenomenon of organismal agency; an object theory just won't do.

---

[16] The distinction between those things that agents *do* and those things that happen to them is initially due to Anscombe (1957).

10.3.2   *Autonomy*

In constituting their affordances through self-maintaining, self-regulating activities, agents forge for themselves a degree of freedom from the vicissitudes of their environments. In doing so they determine which of the conditions is salient, and what they afford. And they set themselves up to exploit the opportunities those conditions have to offer. This is the sense in which agents are 'autonomous':

> [I]n an autonomous system, the constituent processes (i) recursively depend on each other for their generation and their realization as a network, (ii) constitute the system as a unity in whatever domain they exist, and (iii) determine a domain of possible interactions with the environment. (Thompson 2007: 44)[17]

An autonomous agent is thus capable of making 'sense' of its circumstances (Di Paolo 2005; Thompson 2007). The interpretation we apply to 'making sense' here is significant. 'Making' is meant literally here. Making is not merely detecting the features of one's situation, it is in part constituting them. An organism 'makes' its affordances by its ability to respond to its conditions. By 'sense' is meant the capacity of an agent to mobilise its resources in a way that is appropriate to the pursuit of its goals, by exploiting the opportunities, or by ameliorating the impediments. An agent makes a feature significant by the way it detects it and responds to it in the pursuit of its goals. In this way, autonomous agents construct (or constitute) the conditions to which they respond.

An agent-centred theory of evolution may appeal. But adopting it engenders a fairly radical revision to the ontology of orthodox evolutionary theory. It requires positing agents as autonomous, purposive systems. It further necessitates the recognition of affordances, as emergent properties, constituted jointly by an organism's conditions of existence, and its capacity to respond to them. It implies the reciprocity of form and affordance. As form evolves, so do the affordances to which it responds.

10.4   Agential emergentism

In positing organismal agents as primitive, Situated Darwinism takes on a commitment to emergence. The dynamics of organisms are to be explained in terms of their purposes and affordances, and these are emergent properties. They are properties of the relation between an agent and its circumstances, but not properties of the system's parts. As we discussed in Chapter 1, mechanism, the methodology of the Modern Synthesis, appears to be inimical to emergence. Mechanism tells us that the dynamics of complex entities can be wholly

---

[17] I thank Alex Djedovic for drawing my attention to this passage.

explained by adverting to their parts. If the properties of complex entities are wholly accounted for by the properties of the parts, then there is no unexplained residuum for the properties of complex entities to explain. Emergent properties are explanatorily redundant, and hence dispensable. But if agency and affordances are genuinely emergent properties, and theoretically indispensable, that, in turn, demands that emergent properties of a complex entity – in our case, agency – really can explain some features not wholly accounted for by the several properties of the parts. If we are helping ourselves to indispensable emergent properties, it also behooves us to address the widely held conviction that the very idea of an emergent property is somehow incoherent. I take these in turn.

10.4.1   Downward explanation

Emergence ought to hold no fears for contemporary science.[18] There is a considerable amount of explanatory work that emergent properties can do. Schrödinger's order-from-disorder answer to his '*What is Life?*' question offers us a hint at how a theory that incorporates emergent properties of complex entities might work.

Far-from thermodynamic equilibrium systems spontaneously build structures that dissipate energy. As they do, they become increasingly stable, and they complexify. Such systems have their own intrinsic dynamics. The dynamics of these systems can yield predictions and explanations, not just about the activities of the system as a whole but also about the activities of the parts. The fact that a water molecule is part of a Bénard cell, may tell us more about its future trajectory, than its current velocity does, or for that matter the relations between its atomic parts. The dynamics of convection confers on our target molecule a trajectory that it would not otherwise have. Of course, we can explain the dynamics of a Bénard cell by aggregating the motions of its individual water molecules. But, we can also explain the motions of an individual molecule by citing the dynamics of the cell of which it is a part. There is a sort of explanatory reciprocity between parts and wholes in self-organising systems. In addition to the bottom-up explanations of mechanism, there are top-down 'emergent' explanations that advert to the dynamics of complex systems (Davies 2012; Mitchell 2012; Walsh 2012b).

Organisms are extreme cases of such self-organising, self-building, self-complexifying systems. I have characterised this dynamics as 'purposive'. This purposiveness figures in explanations of the activities of the entire systems as well as explanations of the structures and activities of the organism's parts.

---

[18] Deacon (2011) offers an extended argument in favour of emergence, one that diverges significantly from my own.

Organisms have the capacity to construct their own parts, and to enlist and regulate the causal capacities of their parts, in such a way as to ensure the pursuit of their goals.

Situated Darwinism stresses the importance of these top-down explanations for understanding the process of evolution. The activities of individual genes, genomes, cells, tissues, organs and immune, thermoregulatory, metabolic, behavioural and cognitive systems can be explained in large measure by adverting to the purposive dynamics of the organisms of which they are a part. Organisms must make their parts work in ways that are conducive to the fulfilment of their goals. So one of the ways that organisms constitute and exploit their conditions of existence is by regulating the causal powers of their component parts (Walsh 2013a).

Granted, as mechanists insist, organisms inherit their causal capacities from their parts. And so the capacities of whole organisms can be accounted for from the bottom up. But that is only half of the explanatory picture. The parts have their particular capacities, in these instances, because of the systems in which they are embedded. There is a reciprocal relation of regulation between the activities of an organism's parts, and the activities of the organism as a whole. Organisms are truly, in Kant's felicitous phrase, 'both causes and effects of themselves' (Walsh 2007b).

An overreliance on the bottom-up strategy of mechanism obscures the explanatory reciprocity between parts and wholes. It gives us organisms as effects of their parts, but not organisms as causes of the existence and the activities of their parts. Mechanism can tell us why, given the causal capacities of an organism's parts, organisms behave in the way that they do. But it cannot answer the analogue of Smolin's question: 'why do the parts have those particular causal powers rather than some other?'. The parts of organisms get their particular causal powers from their contexts – the activities of the organisms as a whole – just as a water molecule gets its movements from the Bénard cell of which it is a part.

10.4.2   Reflexive downward causation

The control that organisms exert over the causal powers of their parts is an instance of what Jaegwon Kim calls 'reflexive downward causation' (Kim 1999, 2006). As we saw in Chapter 1, Kim thinks the very idea is incoherent. There, I quoted him saying:

Emergentism cannot live with downward causation and it cannot live without it. Downward causation is the raison d'etre of emergence, but it may well turn out to be what in the end undermines it. (Kim 2006: 548)

Were that so, it would be fatal to Situated Darwinism. So, Kim's antipathy to reflexive downward causation needs to be addressed.

Kim's argument is predicated on two assumptions. The first he makes explicitly, the other only implicitly. The explicit assumption is the Causal Inheritance Principle. It says that the causal powers of a complex system are inherited exclusively from the causal powers of their parts. Causal inheritance has two putative implications. The first is that were the parts of a given complex system not to have their causal capacities, then the system as a whole wouldn't have its causal capacities. The capacities of the whole counterfactually depend upon the capacities of the parts. That seems entirely reasonable.[19] The second is that, for a complex entity, nothing other than its parts is relevant to the determination of its causal properties. That in turn requires that the causal powers of an entity are intrinsic to it. And *that* is the implicit assumption.

The assumption of the intrinsicality of causal powers is a compelling one. Since Kripke (1972) and Putnam (1970) philosophers have become accustomed to thinking that what makes an entity the kind of thing it is – what determines its characteristic activities – is an intrinsic causal disposition (Bird 2007). Paradigmatically, gold is what it is, and does what it does, because of its intrinsic, dispositional causal powers. These, in turn, are imparted to atoms of gold by their structure.[20] Intrinsic properties are context insensitive. An entity has all its intrinsic properties (at least until it undergoes an internal change), irrespective of the context that it is in. If causal powers are intrinsic, nothing other than the internal constitution of the entity itself could confer on it its causal powers.

The Causal Inheritance Principle conjoined with the assumption of the intrinsicality of causal properties confers an ontological priority on the capacities of parts over the capacities of their aggregates. Complex entities can (indeed must) inherit their causal powers from their parts, but the converse relation does not (could not) hold. A complex entity could not confer on its parts causal powers that the parts did not have by their own intrinsic natures. Consequently, the properties of a complex entity could not explain why its parts have their causal powers. That is the rationale for Kim's stricture against reflexive downward causation. The presumed intrinsicality of causal properties is crucial to Kim's case against emergence.

Perhaps the idea that causal powers are intrinsic derives from the intuition that fundamental causal powers, like mass for example, are context insensitive. The mass of a macroscopic object is not altered by the masses of other bodies. Mass behaves in a context-insensitive way with respect to forces. This is extremely useful, of course. The mass of an object allows us to predict its behaviour across a range of contexts in which various forces are acting upon it.

---

[19] This is the principal that lends modern mechanism its air of reductionism: the capacities and activities of complex entities are fixed by the capacities and activities of their parts.

[20] See Ellis (2007) for a sophisticated treatment of dispositional essentialism.

It allows us to decompose the several forces acting on a body, as it allows us to assume their effects are mutually independent, and that none of them affects the body's mass. The context insensitivity of causal powers appears to be an integral piece of what Nancy Cartwright calls the 'analytic method'. Here it is assumed that as contexts change, an entity's causal powers remain unaltered. The intrinsicality of causal powers secured their constancy across contexts.[21]

But it is important to note that context-insensitivity of this sort, does not entail intrinsicality. Mass itself provides a vivid illustration. Mass may be invariant across the range of contexts with which we usually deal, but it is not an intrinsic property of a body. The mass of a particle appears to be an 'emergent' property (Smolin 2013). It arises out of the interaction of a particle with the Higgs field. Mass is 'conferred' upon a particle by its relation to something else. So, causal powers, even if they are invariant across a wide range of contexts, can be relational, properties of things.[22]

Where causal powers are nonintrinsic properties, conferred on things by their context, it is perfectly coherent to suppose that the parts of a complex system get their causal powers from the system as a whole. The parts would not have had those particular capacities were they not parts of that particular complex entity. The whole system is the context that confers on the parts their causal powers. This holds *even if* the causal powers of the system as a whole are entirely inherited from its parts. For example, a particular gene may have a particular causal role, say the regulation of another gene product, only in the context of the gene regulatory network of which it is a component. Nevertheless, that gene regulatory network consists entirely of the relations between its constituent genes. We have a reciprocal kind of counterfactual dependence: the capacities of complex systems counterfactually depend upon the capacities of their parts, and the capacities of the parts counterfactually depend upon the properties of the whole.

The reciprocal dependence of the properties of parts and their wholes sounds a little paradoxical. I suspect the oddness derives from an implicit tendency to think of the properties in question – causal powers – as intrinsic. But this is not at all strange when we allow the relevant properties of the parts to be context-sensitive and relational. Consider an example. Charles and Emma are a (married) couple. Their being a couple consists wholly in the relation that each bears to the other – Charles's being married to Emma, and Emma's being married to Charles. Yet the relational property that each partner has of being married to the other is conferred on each by the fact of their being a married couple. So, the

---

[21] This is the foundational supposition of modern mechanism. The capacities of Boyle's 'atoms' and Descartes' 'corpuscles', for example, are intrinsic and context-sensitive. They will '... in each case, act in accordance with their nature' (Cartwright 1999: 83). See the discussion in Chapter 1.

[22] See Walsh (1999) for a discussion of context-sensitive, relational causal powers.

property of the whole (*connubiality*) depends upon the properties of the parts (in this case: *being married to Charles* and *being married to Emma*), and reciprocally, the properties of the parts depend upon the properties of the whole.

Similarly, where causal powers are permitted to be relational properties, the possibility arises that complex systems may have the powers they have solely in virtue of the causal powers of their parts. (This is 'Causal Inheritance'.) Yet at the same time the parts have their relevant causal powers in virtue of being parts of the complex system in question. This is the dreaded 'Reflexive Downward Causation', and it is far from incoherent. It is especially plausible in causally cyclical systems to suppose that the relevant causal powers of the parts of the system are highly context sensitive and conferred on them by the system of which they are parts.

If this is the right way to think of emergence, then emergence simply consists in the fact that complex systems transform the capacities of their parts (Ganeri 2011). They confer on their parts capacities that they would not otherwise have. These in turn fix the properties of the complex system.[23]

Emergence, conceived in this way, is an extremely banal, workaday phenomenon. It poses neither threat nor puzzle. It arises from the fact that causal powers can be context dependent and relational, and that the system of which an entity is a part can be a relevant context for the determination of causal powers. In extremely complex systems, like organisms, however, it has some quite remarkable consequences. It demonstrates the Kantian reciprocity between the causal powers of the whole and those of its parts that is diagnostic of complex adaptive systems. It further suggests an explanatory reciprocity that cannot be countenanced by standard mechanism. The causal powers of the parts can be explained in terms of the properties of the system as a whole, even if the capacities of the system as a whole are also explained by appeal to the causal powers of the parts.

Something like this is probably all that the British Emergentists had in mind. In particular, it seems to be clearly reflected in Mills' claim that:

> ... the phenomena of life which result from the juxtaposition of those parts in a certain manner, bear no analogy to any of the effects which would be produced by the action of the component substances considered as mere physical agents. (Mill 1843: 243)

Emergence may violate the precepts of mechanism, but it is in no way incoherent. It is a wholly natural, unsurprising, eminently observable phenomenon. The British Emergentist, Samuel Alexander captures the spirit succinctly.

---

[23] At least they contribute to the properties of the complex system. Where causal properties are allowed to be context sensitive, we may suppose that complex systems pick up causal properties from their external relations too.

The existence of emergent qualities thus described is something to be noted, as some would say, under the compulsion of brute empirical fact, or, as I should say in less harsh terms, to be accepted with the 'natural piety' of the investigator. (Alexander 1920: 46–47)

There should be no compelling objections, then, at least on metaphysical grounds, to accepting an ontology for evolutionary biology in which organisms as agents confer on their component parts their distinctive capacities. Nor should there be any particular metaphysical scruples about taking evolution to be the consequence of the response by agents to their affordances.

10.5   The disappearing environment

Having salvaged the organism, Situated Darwinism, as an agent theory, runs headlong into the obverse problem: the disappearing environment, as a 'given'. The environment has a special theoretical significance for orthodox Modern Synthesis theory. It is necessary for explaining what makes evolution adaptive. The idea is that natural selection promotes those traits (or individuals) that possess higher fitness. For that process to eventuate in the increased adaptedness, higher fitness must correspond with greater adaptedness. 'Adaptedness' in this sense is aptness to survive in the external environment. So, for greater fitness to correspond to greater adaptedness, there must be a common ground against which fitnesses can be measured. That common ground is generally thought to be what Robert Brandon calls the 'selective environment' (Brandon 1990).

The selective environment is the critical notion for evolutionary theory, because it is what organisms must share if they are to be subject to a single selection process . . . .' (Sterelny and Griffith 1999: 269)

In Situated Darwinism's agent theory, the environment does not appear as an unmediated cause of evolution. There is to be sure an external environment, and it does have causal efficacy. But, as we discussed in Chapter 8, the external environment radically underdetermines the challenges posed to biological form, and hence underdetermines what would count as an adequate adaptive response. Consequently, it is not the environment *per se* that explains evolutionary changes in form; it is the system of affordances.

The system of affordances is quite unlike the environment. Not only is the system of affordances not external to the organisms in a population, it isn't shared by them either. Jacob von Uexküll captures the idea vividly. 'Every animal is surrounded by different things, the dog is surrounded by dog things and the dragonfly is surrounded by dragonfly things' (von Uexküll 1938: 117).[24] In fact, affordance systems are even more finely individuated than

---

[24] I first encountered this quotation in Van Duin, Keijzer and Franken (2006).

even von Uexküll suggests. An organism's complete set of affordances is likely to be idiosyncratic. Organisms vary. Each individual's set of features and capacities is different. Moreover, the physical conditions that each organism encounters are minutely different. So, there is no set of conditions that *both* causes the survival and reproduction of organisms *and* is held in common amongst all the individuals in a population at a time.[25] Positing organisms as primitive agents in evolution has led to the disappearing environment.

The concern is that without positing environments as shared external, causal factors, the theory of evolution cannot account for how natural selection leads to adaptive population change. This, as we saw in Chapter 8, is the nub of Lewontin's worry. Certainly, Darwin's theory of adaptive population change appears to require the existence of objective, external 'stations' in the economy of nature. The disappearing environment could be just as detrimental to Situated Darwinism as the disappearing organism is for the Modern Synthesis. So, it is worth asking whether evolutionary theory could survive without positing a shared external physical environment as basic.

10.5.1   The importance of the environment

Some authors suppose that a common environment is a precondition for the applicability of population dynamics models (Millstein 2006; Otsuka et al. 2011). Unless a group of organisms shares a common environment, they do not constitute a biological population. Consequently, there is no way that we can apply the formal models of population dynamics to them. This line of thought strikes me as naive and implausible. It is worth bearing in mind that the models of population dynamics, of the sort initiated by Fisher, have no term for a shared environment.[26] These models simply assign growth rates – fitnesses – to abstract trait types. As intrinsic growth rates, trait fitnesses are not relations to environments, in the way that individual fitnesses (arguably) are. The models demonstrate what happens to an ensemble subdivided into subclasses in which growth rates among the subclasses differ.

In fact, it bears stressing that these models, as abstract as they are, have no strictly biological content. They can be applied equally to changes in the structure of any number of kinds of ensembles, from trait types to investment portfolios (Matthen and Ariew 2002; Orr 2007b; Walsh 2014a). As these are not 'biological' models *per se*, there are no specifically biological conditions that their application must meet. In particular, the use of these models to represent the change in abstract trait structure of a population does not require

---

[25] I thank Lynn Chiu for pressing this problem.
[26] Fisher explicitly disavowed Darwin's 'Malthusian' assumption of an environment holding populations in check.

a shared, external environment as the cause of differential trait growth (Ariew and Ernst 2009).

The claim that the models of population dynamics require that the systems they apply to share an environment seems to have fallen prey to an equivocation on the concept of a population. A biological population is a collection of conspecific organisms living in relative proximity, causally interacting with one another. It is reasonable to suppose that such collections must share some features of a physical environment (Stegenga 2014; Winther 2015). A population as it figures in evolutionary population dynamics is different. It is a collection of abstract types; any system to which these models can be usefully applied qualifies as a 'population', in this sense. Winther et al. (2015) propose the terms 'natural population' and 'theoretical population' (respectively) for these entities. Theoretical populations are the objects of population dynamics models. Their members do not need to share an environment.

If a shared causal environment is not a requirement of the application of quantitative models of population dynamics, and it is not the cause of adaptive evolution, what is its importance to evolutionary theory? One intriguing idea is that it plays something like a heuristic role. Trevor Pearce (2010) notes that the concept the environment as a self-standing, unified causal entity is an invention, and a fairly recent one at that. It postdates Darwin's theory of natural selection.[27]

In terms of metaphysics, the successive transitions from individuated particular factors (e.g., climate), to a general plural term (e.g., 'circumstances'), to a general singular term (e.g., 'environment'), correspond to a progressive concealment of the different elements that make up the world outside the organism and the relations between these elements. This concealment, perhaps misleadingly, implies that the environment can be taken to be a single, unified cause. (Pearce 2010: 249)

The shared external environment is a construct. It is cobbled together out of the enormously complicated constellation of causal relations that each individual organism encounters. It is a pragmatic or heuristic device for applying our models generally to biological populations. Even though each individual organism occupies an idiosyncratic system of affordances, we can generate the shared environment as a sort of simplifying approximation, or an idealisation.[28]

Each organism's affordances are unique, yet similar organisms experience similar affordances in similar conditions. As environments partially constitute those affordances, the affordances of similar organisms will correspond, to

---

[27] Pearce traces its inception to Spencer.
[28] Similar points are made by Kaplan (2013) and Matthen (2009). I thank Jonathan Kaplan for drawing my attention to this similarity.

some reasonable degree, to their shared environments. By abstracting away from the multiple differences between individuals, and by idealising away the minute variations in external conditions, we can generate the idea of 'the environment'. 'The environment' comes to stand as an index for the conditions that any arbitrarily chosen member of a population is likely to encounter. This in turn can be used as a basis for assigning, and projecting, comparative fitnesses. It also becomes a useful device for helping us understand the functions of adaptive structures and processes.

As an idealisation, the shared environment doesn't need to be the cause of evolution *per se*, in order to fulfil its theoretical role. Trevor Pearce continues (from above):

However, the singular term 'environment', . . . , is an important heuristic for biologists, insofar as it gives them a way to talk about general causes without exploring the details of micro-level complexity (the term 'natural selection' is a parallel case). (Pearce 2010: 249)

The 'environment' is a device for making inordinate complexity of biological reality a little more tractable.

10.5.2   Models and the environment

In his landmark paper 'The Strategy of Model Building in Population Biology', Richard Levins (1966) observes that intractable complexity is a general feature of the real-word systems that we wish our scientific models to represent. Many systems have: '. . . too many parameters to measure; some are still only vaguely defined; many would require a lifetime each for their measurement' (Levins 1966: 18). We could not possibly represent all the causally relevant details of evolution in a model. Nor would we especially want to. Such a model would be unwieldy, incomprehensible and unillumi-nating. So we must simplify.

Streamlining our models involves making choices about what to represent explicitly, what to leave out, what to approximate, and what to fudge (Walsh 2014d). As Levins points out, those choices depend upon our specific objec-tives. In general, trade-offs must be made between three potentially good-making virtues of a model: realism, precision and generality (Weisberg 2006). Levins enumerates various strategies that we might follow in construct-ing our models. One, in particular, is relevant to models of evolutionary population dynamics.

Sacrificing realism to gain generality and precision is the second strategy of model building. If you are willing to make a large number of approximations and work with a highly abstract representation, you can generate precisely specified models that will apply approximately to many target systems. (Weisberg 2006: 638)

Population dynamics models do just this. They are applicable across an enormous range of disparate kinds of biological populations (Ariew 2008). If fitnesses are apportioned correctly, a model can generate powerful explanations, and robust predictions.

But having traded off generality and precision against realism, we must be circumspect in drawing conclusions concerning what these models say about the *real* workings of the biological world. The relation between the models and their real-world object systems is in no way transparent. The models carry no assurances that they represent the metaphysics of population change with any great measure of fidelity. Like fables and parables, a considerable amount of extra work is needed to interpret them (Cartwright 2010). So, if a model of adaptive evolution represents organisms (or traits) as possessing fitnesses relative to a shared or common environment, we are not thereby licensed to infer that there is a unified entity, 'the common environment', or that it is exerting any significant or isolable causal influence on the evolution of form.

An abstract notion of the 'shared environment' can capture something of importance to the explanation of adaptive population change. It can represent the fact that in very roughly constant conditions, the changes in population structure explained by differential fitnesses can realise adaptive evolution.[29] But it does not require us to hypostatise environments as autonomous causal entities. They are simply idealised parameters in a highly abstract, simplified model.[30]

Situated Darwinism is a thesis about the metaphysics of evolutionary change. It says that evolution is a consequence of agency – organisms' purposive engagement with their affordances. Affordances are not shared environments. And shared environments are not unmediated causes of adaptive population change. Nevertheless, Modern Synthesis models of population change can measure fitnesses relative to abstracted, idealised common environment. And they can do their theoretical work with remarkable potency. So long as we do not interpret these models literally – as committing to shared environments as autonomous causes of change – we are free to use Modern Synthesis models as representations of the effect on populations (of trait types) that accrues from organisms as agents engaging with their affordances.

10.6   Modelling and metaphysics

I began this chapter by pointing to what looks like a deep incompatibility between the Modern Synthesis theory and Situated Darwinism. As an agent theory, the latter takes evolution to be the consequence of organisms as

---

[29] Of course, more conditions are required (Kauffman 1993). See Chapter 7.
[30] Similar considerations apply to the 'genes' represented in these models.

purposive agents responding to and creating their conditions of existence. As an object theory, the former takes evolution to be the consequence of external causes – selection and drift, the influence of the environment – acting upon genes. On one view, organisms are agents that enact evolution. On the other, they are passive objects of evolutionary forces.

The foregoing discussion of the role of the environment, however, suggests a possible of reconciliation. The Modern Synthesis theory is an extremely powerful tool for representing the dynamics of populations. Its models are highly abstract. Their ontologies typically involve abstract trait types, and not concrete entities. Changes in the relative frequencies of these abstract types are explained and projected by the variation in their fitness. But, as we suggested in Chapter 2, the power of these models comes at a cost. By abstracting from the biological facts on the ground in this way, these models forfeit their claim to carving the process of evolution at its joints. Perhaps, then, we should not expect the models generated under the auspices of the Modern Synthesis to tell us anything substantive or reliable about the metaphysics of evolution.

Situated Darwinism may provide a more satisfactory account of the metaphysics of evolution. In particular, it gives us a more realistic understanding of why biological evolution should be adaptive. It adequately accounts for the way that the features that distinguish organisms contribute to the conditions required for evolution. It correctly identifies the place of genes, organisms, and environments in evolution. But it too has its drawbacks. Significantly, it isn't so obviously clear how to generate from it any usable, generalised and accurate models of evolutionary change.

The proposed concordat resides in the idea that while the Modern Synthesis theory gives us a way of modelling evolutionary change in population structure, Situated Darwinism offers an account of what happens when biological evolution happens. The process that the abstract Modern Synthesis population models represent is not the process by which extrinsic forces or causal processes – selection and drift – mould biological form by pushing abstract gene types around genotype space by dint of their fitnesses. These parameters, selection, drift, fitness, do not figure intimately in the process of evolution, although they are indispensable to the models that describe it. The process that these models represent is one in which organisms engage with, and construct, their affordances. None of these phenomena – purposes, affordances, agency – is explicitly represented in the models. But they are nevertheless real. The Modern Synthesis and Situated Darwinism tell us different things about evolution. Modern Synthesis models tell us about the evolutionary dynamics of populations. Situated Darwinism tells us about the metaphysics of evolution. They are clearly compatible if we recognise the division of epistemic labour. Modern Synthesis models tell us about the evolutionary dynamics of populations; Situated Darwinism tells us about the metaphysics of evolution. They only appear incompatible if we ask them to do the same work.

Conclusion

Organisms as agents were omitted from the Modern Synthesis, but not because someone forgot to put them in. They were left out because the Modern Synthesis is an object theory, it has no way to accommodate agents. One cannot simply graft the idea of organisms as agents onto Modern Synthesis evolutionary biology, while leaving the rest of the theoretical apparatus intact. Consequently, an evolutionary theory that deals in agents, purpose and affordances is no mere extension of the Modern Synthesis; it is 'a different *kind* of theory'. As an agent theory of evolutionary change, Situated Darwinism seeks a place for the reciprocity that holds between biological form and the conditions under which it evolves. In dealing with agency, purposes and affordances, it is heavily invested in emergence. All these commitments are anathema to a mechanistic object theory. It thus seems hard to square Situated Darwinism with Modern Synthesis evolution. One possible line of reconciliation is to suggest that Situated Darwinism renders a realistic account of the metaphysics of evolution, while the Modern Synthesis provides a set of extremely useful, albeit abstract and general models of evolutionary change.

11 Two neo-Darwinisms
Fractionated or situated?

The Modern Synthesis theory of evolution and Situated Darwinism share a common ancestor. They are both extensions of the theory of fit and diversity of form adumbrated in *The Origin of Species*. Only part of the family history has been told, however. The genealogy of the Modern Synthesis has been meticulously traced (Cain 2009; Depew and Weber 1995; Mayr and Provine 1980; Provine 1971). Its growth and change have been documented in minute detail. The story of the other line of descent, quite understandably, has yet to be related.

The changes that occurred in evolutionary theory in the transition to the Modern Synthesis are sometimes regarded as exclusively empirical in nature. They are simply the consequence of filling in the lacunae in our biological knowledge. In the process, it is usually supposed, the core of Darwin's theory has been left undisturbed. It might seem obvious and unavoidable that as our understanding of the mechanisms at work within organisms grows – particularly our understanding of the role of genes in transmission and development – the explanatory focus of evolutionary theory should shift from the activities of organisms to the activities of the suborganismal entities that constitute them. And in concentrating our attentions on the suborganismal realm, so the story goes, there is substantial empirical gain at no theoretical cost. The gene's-eye view of evolution is simply *The Origin* brought into sharper focus – Darwinism in detail.

This revisionist reconstruction underestimates the divergence between Darwin's theory and its Modern Synthesis successor. In turn, it fosters the impression that this is the only way that evolutionary thinking might have evolved. Indeed, throughout most of the twentieth century, Modern Synthesis theory prosecuted an uncontested claim as the true heir of Darwinism. Perhaps, in the absence of a viable pretender, this is a reasonable supposition. But, to my knowledge, primogeniture is not a principle of theory choice. If there are multiple claimants to Darwinism's mantle, they must stand or fall on their own merits.

The proposed alternative, Situated Darwinism, as its name is meant to convey, also traces its roots back to the *Origin of Species*. In particular, it accentuates the very feature of Darwin's theory that the Modern Synthesis neglects, the role of organisms. Situated Darwinism expands on Darwin's crucial notion of an organism's 'struggle for existence'. The 'struggle' consists in the capacity of organisms as purposive systems both to construct and to respond to their affordances. Evolution is the historical trace that the struggle leaves behind.

There is common descent to be sure, but there is modification, too. Modern Synthesis thinking and Situated Darwinism differ in their explanatory structure, and each departs from their common ancestor. They differ in their canonical unit of biological organisation. They differ in their characterisations of the quintessentially biological phenomena of inheritance, development and adaptive change. They differ in the interpretation each gives to the nature of selection and the role of chance. Crucially, there are evolutionary events countenanced and explained by Situated Darwinism that do not qualify as evolution on the Modern Synthesis replicator theory. Specifically, changes in phenotype – form – that are not accompanied by changes in genotype space do not count as evolutionary changes according to Modern Synthesis replicator theory, but may do so from the perspective of Situated Darwinism.

They differ in their standing. The Modern Synthesis is the established theory of evolution. It enjoys the cachet that attends a theory in a period of normal science, and it boasts a century of success. Recently, however, there have been susurrations of unrest (Laland et al. 2014; Noble 2006; Shapiro 2011; Wray et al. 2014). For the reasons we surveyed in Part II, there are increasingly frequent and strident calls for the reassessment, extension (Pigliucci 2009a, 2009b, 2010), and wholesale revision of the Modern Synthesis (Laubichler 2010; Maienschein and Laubichler 2014). Some critics have even been emboldened to declare that 'Darwinism in its current scientific incarnation has pretty much reached the end of its rope' (Depew and Weber 2011: 90). That may or may not be so, but the current uncertainty does at least provide some impetus to the consideration of an alternative.

By comparison, Situated Darwinism is nascent, inchoate, struggling for a definitive articulation. No doubt it will take a considerable amount of time to grow, and it will change along the way, if it survives its infancy at all. Nevertheless, it is possible to foresee in rough outline how this alternative conception of evolutionary biology might develop. Luckily the materials are mostly already to hand, having been gathered over the course of the preceding chapters. In what follows I shall attempt to expand upon the leading ideas in Situated Darwinism, largely by contrast, using Modern Synthesis thinking as a foil. The principal difference, as I have been stressing, is that Situated Darwinism takes evolution to result from the agency of organisms. I shall then survey some of the ways that the purposive activities of organisms contribute to evolution.

11.1   Evolution as an ecological phenomenon

The central positive proposal running through the preceding chapters is that evolution is an 'ecological' phenomenon, in contrast to the 'molecular' process it is depicted as being by the Modern Synthesis. By characterising evolution in this way, I mean to emphasise that it is a kind of by-product of organisms as agents doing what they do. Organisms struggle and as a consequence populations (and lineages) evolve. The struggle consists in the purposive engagement of organisms with their affordances. Evolution may be registered as a change in the gene structure of a population, but unlike Modern Synthesis thinking, Situated Darwinism holds that that is not its essence.

11.1.1   The Loss of Distinctions

Modern Synthesis evolutionary thinking is predicated upon a series of fundamental distinctions: between the component processes of evolution, between evolutionary and nonevolutionary characters, between the external and internal influences on form. Situated Darwinism, for its part suggests the dissolution of these.

Inheritance, development, and selection   In Chapter 3 I stressed that the core of the Modern Synthesis is a commitment to fractionation. The Modern Synthesis crucially represents the component processes of evolution – inheritance, development, adaptive population, and the origin of novelties – as quasi-independent, each with its own proprietary cause. Inheritance is the transmission of replicated entities from one generation to the next. Development is the expression of the phenotypic information encoded in the replicators. Together, inheritance and development secure the conservativeness of form upon which evolution depends. These 'conservative' processes are wholly separate from the processes that promote evolutionary change: mutation, and selection. The presumptive division of causal labour is appealing and convenient; it induces a nice, clean division of explanatory labour. But given what we now know about the processes occurring within individual organisms, fractionation is becoming increasingly hard to sustain.

It is of the utmost importance for Modern Synthesis thinking that there is a further mutual independence between the processes that underwrite inheritance and development. One reason for this is a presumed asymmetry between inheritance and development: organisms develop the traits they inherit, but do not typically inherit the traits they merely develop.

The independence of inheritance and development supports what has long been thought to be a crucial distinction between 'nature' and 'nurture'. Some of our character traits are the legacy of our ancestors, bequeathed to us, unalterable and inevitable. They constitute a blueprint of what we are, of our innate capacities. Other characters are merely an overlay on this inherited plan. These are the acquired characters, and they play no role in the evolution of lineages. These distinctions are often thought to be misguided or pernicious (Bateson and Mameli 2007; Keller 2011b; Lewontin 1983; Oyama 1985, 2000). But, as Keller (2011b) so elegantly documents, the conviction that there is a genuine difference here, is extremely difficult to dislodge. The conviction may be hard to shake off in part because some version of it is written into the constitution of the Modern Synthesis.

If inheritance and development introduce no systematic adaptive bias, another *further* process is needed that does. Selection imposes changes in phenotype space by choosing amongst those individuals better suited to their conditions of existence. As selection is wholly distinct from inheritance and development, it requires a cause that is distinct from those processes occurring as well. The usual supposition is that the source of selection is the environment external to the organism. In orthodox Modern Synthesis thinking, then, in order for selection to be distinguishable from inheritance and development, the agent of selection – the environment – must be self-standing and autonomous, external to organisms, and capable of exerting its causal influence unmediated by organisms themselves.

The discreteness of selection introduces a further distinction that is pivotal to Modern Synthesis thinking, this one between kinds of explanation. As inheritance, mutation and development are 'internal' processes, occurring within organisms, and selection is a process imposed on form by the external conditions of existence, there are two kinds of causal explanations in evolution: internalist and externalist. Explanations of the ways that inheritance and development constrain evolutionary change advert to internal processes. They are internalist. Explanations that advert to the power of selection to modify form adaptively, appeal to the capacity of the shared external environment to choose between organisms. They are externalist.

It is this schema that pits 'form' against 'function' (Amundson 2005), development against selection, the forces of conservatism against the forces of change. This same schema relegates organismal development to a place of minor importance in evolution. As development neither introduces, transmits, nor chooses between evolutionary characters, it is the least significant of the four component processes of evolution. It is merely the processes by which replicators are delivered to the tribunal of the environment.

Situated Darwinism inverts this priority. It takes organismal development (broadly construed) to be the most significant component process in evolution. Development here encompasses any of the processes that contributes to the formation, maintenance or alteration of an individual organism's form, function, or its interaction with its conditions of existence. It isn't merely the translation of a genetic code into phenotype. The development of organisms, furthermore, is the ultimate source of evolutionary novelties; it underwrites the transgenerational stability of form necessary for inheritance, and it biases evolutionary change.

Situated Darwinism not only inverts the priority of the component processes of evolution, it reconfigures them. In particular it dissolves the traditional distinctions that form the cornerstone of Modern Synthesis thinking.

Inheritance and development   The inheritance that is required for evolution consists in a pattern of resemblance and difference between organisms. There must be intergenerational stability of *intra*lineage similarities and *inter*lineage differences. Roughly speaking, offspring should resemble their parents more than they resemble arbitrarily chosen members of their parents' generation. If that is the sense in which inheritance is required for evolution, it must be acknowledged that this pattern is held in place by more than just the transmission of replicators. To be sure, the actions of gene replication, transcription and regulation are integral to the pattern of inheritance, but so too are the adaptive responses of the genome, the cell and the entire organism. The organism, as an adaptive, purposive agent, assimilates, transduces, harmonises and orchestrates these myriad causal influences. In doing so it secures the differential resemblance of offspring to their parents. The intergenerational stability of form is thus a consequence of the robustness of development broadly construed.[1]

The participation of development in inheritance has three implications. The first is that the processes of inheritance and development are not discrete and autonomous. In fact, contrary to the Modern Synthesis view, inheritance is not so much a process as an effect. It is the *pattern* of resemblance and difference that results from the process of development. That, at least, is how Darwin's theory of natural selection (as distinct from his theory of pangenesis) treats it. The transmission of replicators participates in inheritance. But it is neither necessary nor sufficient to secure the pattern of resemblance and difference that constitutes inheritance. That much has been repeatedly demonstrated by inheritance pluralism and Developmental Systems Theory.

The second implication concerns the putative distinction between innate and acquired characters. Under Situated Darwinism it simply disappears. There may be a difference in the degree of robustness between characters. Some traits are canalised (Ariew 1996, 1999), or developmentally (generatively) entrenched (Wimsatt 2000). These traits may be more reliably produced generation-on-generation. But they are not different in kind from less robustly produced, less reliably recurring traits.

---

[1] I thank Cory Lewis for help with this passage.

The dissolution of the innate/acquired distinction has sometimes been derided as falling hostage to Lamarckian inheritance. In the past the mere invocation of Lamarck seems to have been sufficient to condemn a view as unscientific (Haig 2011). Lamarckianism, by which I mean the thesis that traits that are acquired during the development of an individual can be inherited, appears preposterous only from the perspective of the Modern Synthesis theory of inheritance, in which the only kind of inheritance is 'hard' inheritance. Increasingly, however, the Lamarckian epithet is being accepted, even embraced, by philosophers and biologists (Jablonka and Lamb 1995).[2] The only real question about the transgenerational stability of extragenetically initiated characters is how significant they are for evolution.

As the certainties of Modern Synthesis thinking give way, so too should its proprietary theoretical definitions of biological concepts. One prime candidate for rejection is the idea that inheritance is 'hard inheritance'. Hard inheritance, as we saw, is the idea that replicators are the only things that are genuinely inherited, and nothing other than replication contributes to inheritance. As evolutionary biology increasingly draws away from the conception of inheritance as the transmission of replicators, so too should its antipathy to inheritance as resemblance.[3] Any intergenerationally stable form should count as potentially inheritable, howsoever its stability is secured. Situated Darwinism suggests that the resources for the maintenance of the intergenerational stability of form are various, and spread throughout the gene-organism-environment system.[4]

Evolutionary versus nonevolutionary characters   It follows that there is no distinction in kind either between evolutionary characters and nonevolutionary characters. Any intergenerationally stable trait may potentially be an evolutionary trait. Any such trait can systematically contribute to the fitness of the individual organisms that possess it (or the fitness of others). It can thus change in relative frequency over time in a predictable way. There may be a difference in the degree to which a trait is subject to evolution, as some traits are more robustly stable than others. But this is not a difference in kind.

The Weismann barrier   The distinction between innate and acquired traits, between evolutionary and nonevolutionary characters, between the inheritable and the noninheritable is marked, in standard Modern Synthesis thinking, by the Weismann barrier. The barrier is, allegedly, a consequence of the fact that changes wrought to developmental systems downstream of DNA

---

[2] See various contributions to Jablonka and Gissis (2011).
[3] That is not to say that it should embrace 'real' Lamarkianism, either.
[4] This, I take it, is a central pillar of Developmental Systems Theory (Oyama 1985, 2000; Oyama, Griffiths and Gray 2001).

do not redound to DNA structure. The Weismann barrier seems to be a sort of bulwark of the Modern Synthesis. Without it, there would be no grounds for attributing theoretical primacy to genes. With it, there seems to be no way that nongenetic materials or processes could contribute in any substantive way to evolution.

Situated Darwinism renders the Weismann barrier an irrelevance. Even *if* the structure of DNA is immune to changes wrought by downstream processes, its *function* isn't (Walsh 2014b). The function of germline entities depends strongly upon their contexts. DNA transcription is regulated, proteins are edited and reformed; gene function modulated by the activities of cells.[5] Besides, DNA itself does not enjoy nearly the degree of nobility commonly attributed to it. It is constantly being edited, rearranged and repaired. Information is written into genomes by the structure of developmental systems. It is increasingly becoming clear that genomes are 'read-write' memory systems, not the read-only memory systems they have traditionally been thought to be (Noble 2013; Shapiro 2011). Cells actively alter the structure and function of genomes.

(i) the concept of an RW [read-write] genome altered by cell action is more compatible with the discoveries of molecular genetics than the pre-DNA idea of a read-only memory (ROM) subject to accidental change; (ii) the concepts of physiological regulation can be applied to the control of the NGE [natural genetic engineering] operators that alter DNA sequences and genome structure in nonrandom and controlled ways. (Shapiro 2014: 2329)

Development and selection   Selection, according to Situated Darwinism is a higher order effect. It is the effect on a population of the effect on individual organisms of the entire suite of 'ecological' causes of birth, survival, reproduction and death. We noted, with Darwin, in Chapter 2 that once these causes are in place, evolution by natural selection, should it occur, occurs spontaneously. There is no need for a further, population-level process to be super-added.

That being so, the bias in evolutionary change must be found exclusively in the processes that dispose individual organisms to survive and reproduce. Indeed, one of the most startling lessons to be drawn from recent developmental biology, discussed in some detail in Chapter 6, is that the plasticity of organismal development is the ultimate source of the systematic bias into evolutionary change. To reiterate: evolution is adaptive because development is adaptive.

The presumed distinction between selection and development enshrined in the Modern Synthesis is a category mistake. To think that development might

---

[5] Griffiths and Stotz (2013) offer a nice catalogue of the various ways.

somehow oppose or constrain selection is to hypostatise an extra process in the world. Selection, in a very real sense, just *is* development (broadly construed). The Modern Synthesis error of hypostatising selection is akin to thinking that the pressure of a volume of gas is something over and above the aggregated momenta of its component molecules. Here again, Modern Synthesis thinking conflates an effect with a causal process, just as it does with inheritance. Selection is what happens to a population as a result of individuals struggling to survive.

Internal and external   The distinction between internalist and externalist explanations is one of the most pervasive features of Modern Synthesis thinking (Amundson 2005; Pigliucci 2009a, 2009b). Modern Synthesis explanations of adaptive evolutionary change are externalist (Godfrey-Smith 2001b; Lewontin 1983). They appeal to a causal agent *external* to organisms as the source of biased evolutionary change, the 'selective environment' (Brandon 1990). Situated Darwinism recognises no such internalist/externalist distinction. Here again, from the perspective of Situated Darwinism, Modern Synthesis thinking commits a metaphysical error. It mistakenly hives off the adaption promoting features of evolutionary change, from the form-preserving features, and ascribes to each a wholly distinct, proprietary proeess. But this division of causal labour does not exist. Development provides both the conservativeness and the mutability of form required for evolution.

Just as adaptive explanations are not 'externalist', developmental explanations are not 'internalist'. They do not limit themselves to processes that occur within organisms. The causes of development, as we discussed in Chapter 7, are distributed throughout the entire genome/organism/environment system. Development is thus not 'internal' but extended. Organisms are thermodynamically open systems, constantly exchanging matter and energy with their environments, building structure, both internally and externally, assimilating various internal and external causes.

There are, of course, influences on form that are external to organisms, and influences that originate internally to organisms. But, according to Situated Darwinism, these do not differ in the kinds of evolutionary effects they produce. Nor, according to Situated Darwinism, can evolutionary changes in form generally be decomposed into those principally caused by internal factors and those principally caused by external factors. As discussed in Chapter 8, there may be a distinction between those changes *initiated* externally from those *initiated* internally, but their *effects* do not differ in kind. Typically, the dynamics of organisms are the consequence of the orchestrated, distributed, densely nonlinear, cyclical causal relations. The contributions of any one factor to the production of form are so intimately commingled with others, that they usually cannot be isolated from that of any other. Consequently the effects on form contributed by the external environment cannot be disentangled or isolated from the effects due to internal influences. The distinction between 'internalist' and externalist' explanations so crucial to the Modern Synthesis, makes little sense from the perspective of Situated Darwinism.

Organism and environment   The organism/environment distinction bears an enormous theoretical burden in Modern Synthesis thinking. It has served twentieth-century biology, and its nineteenth-century precursor, well. Yet, the demise the 'internal/external' distinction places the Modern Synthesis conception of the organism/environment relation under considerable strain. At some point the idea that adaptive evolution can be described as the effect of external environments on passive organismal form must abjured.

Darwin's alienation of the outside from the inside was an absolutely essential step in the development of modern biology. Without it, we would still be wallowing in a mire of obscurantist holism that merged the organic and the inorganic into an unanalyzable whole but the conditions that are necessary for progress at one stage of history become bars to further progress at another. The time has come when further progress in our understanding of nature requires that we reconsider the relationship between the outside and the inside, between organism and environment. (Lewontin 2000: 47)[6]

Situated Darwinism provides the sort of reconsideration that Lewontin is calling for. It recognises no such 'alienation of the inside from the outside'. Of course there are external environments. There are obviously also organisms (and their insides). But, in Situated Darwinism, the organism/environment distinction has virtually no significance for the explanation of the evolution of form. Instead, the agent/affordance relation is theoretically basic; evolutionary thinking starts from there. Affordances are not environments. They are emergent properties of the organism/environment system. They comprise the mutually entangled contributions of both. They cannot usefully be decomposed into organismal and environmental components.

11.1.2   Chance and inherency

Modern Synthesis evolution is 'chance caught on the wing' (Monod 1971). The ultimate source of evolutionary novelties is random genetic mutation. In part, as we saw, this commitment arises from the Modern Synthesis conviction that the only evolutionary process that introduces a bias is natural selection. That being so, whatever process introduces new variants must be unbiased. In part, as we also saw, the conviction arises out of the Modern Synthesis aversion

---

[6] It isn't entirely clear that Darwin's conception of the 'conditions of existence', commits him to any form of externalism. See Pearce (2011) for an illuminating discussion.

to teleology. No process that generates evolutionary novelties can introduce an adaptive bias, lest we be required to explain the origin of those novelties by appeal to the well-being of organisms. According to Situated Darwinism, the driver of evolutionary change is the pursuit of organisms' purposes (given their affordances). In pursuing their distinctive way of life organisms respond adaptively, and some of these adaptive responses initiate new, stable phenotypes, which in turn are maintained and permitted to recur in a population through the adaptive accommodation of development. A good portion of these novelties occur precisely because they are conducive to organismal survival, under the circumstances. The introduction of evolutionary novelties is biased by the purposes of organisms. Sufficient evidence is now available to suggest that the adaptiveness of organisms is a precondition for adaptive evolution. Adaptive evolution is not fundamentally chancy. It is inherent in the purposive activities of organisms.

11.1.3   A criterion of evolution?

The Modern Synthesis provides an iron-clad criterion for evolution. Ever since Dobzhansky's *Genetics and the Origin of Species* (1937), we have become accustomed to thinking that evolution consists exclusively in changes in the gene structure of a population. The advantage to this criterion is that it allows us to differentiate those events that are to count as genuinely evolutionary from those that are not, *at the time of their occurrence*. In turn it permits us to observe and measure evolution occurring in small scales, over short time periods.

It is not obvious, however, that change in gene structure within a population over time is identical to the process that Darwin discovered, for which he adduced such an impressive mass of evidence, and for which he offered his most compelling 'long argument'. That process is the long-term change and stasis of biological form. There are good reasons to believe, of course, that change in gene structure of a population is a component – even a significant component – of evolutionary change. But it is not entirely clear that the two processes are coextensive. But for the Modern Synthesis criterion to be *the* criterion of evolution, they would have to be. It seems to me that the most significant outcome from recent work on inheritance and development is the suggestion that the perfect coincidence between Modern Synthesis evolution and Darwinian evolution is unlikely.

The question whether the Modern Synthesis gives us the criterion of evolution puts recent debates concerning the nature of inheritance, development, plasticity, and their role in evolution in context. These arguments are implicitly, I believe, attempts to cast doubt upon the Modern Synthesis criterion of evolution. They are attempts to point out that there are more processes that contribute to evolution than simply those that directly bring about changes in gene structure.

An example might help. The scenario offered by King (2004) and Newman (2011) of the origin of metazoans is one in which a major and abrupt change in heritable form, multicellularity, arose without an accompanying genetic change. Thereafter, significant evolutionary change ensued. This, most assuredly was an evolutionary event. It led to the enormous radiation and diversification of the animals. But it does not qualify under the Modern Synthesis criterion. The aggregation of unicellular organisms into multicellular organisms was not instigated by any genetic change. Likewise, novel, stable phenotypes initiated by phenotypic plasticity may be evolutionary events. They may lead to significant long-term changes in form, even if they do not themselves constitute a change in gene structure of the population. This is the import of West-Eberhard's dictum that 'genes are followers in evolution, not leaders' (West-Eberhard 2005b: 6543).

It is reasonable to respond on behalf of the Modern Synthesis that *typically* such processes do not lead to evolutionary change, even though some instances might. Inevitably, however, changes in gene structure do lead to evolutionary change.[7] That being so, it would be a mistake to extend the criterion of evolutionary change to include all changes in form that are passive responses to environmental alterations, or changes in form that are the result of organismal plasticity. Such a criterion would be entirely too inclusive and misguided. I agree.

Situated Darwinism suggests a different perspective. As there is no distinction in kind between evolutionary characters and nonevolutionary characters – as there is no difference in kind between evolutionary changes and nonevolutionary changes – we should expect that there is no criterion of evolution. Typically, one cannot say of a change in a population occurring at a time, whether or not it constitutes an evolutionary change. In general, evolutionary events can only be judged to be so in retrospect. The massive agglomeration of single-celled organism may not generally issue in any discernible evolutionary event. But one such historical episode did, and the result was earth-changing. It eventuated in the metazoans. So agglomeration *per se* is neither an evolutionary process nor a nonevolutionary process, but instances of it may be either. The same goes for phenotypic novelties that result from developmental plasticity – I dare say also for genetic mutations.

---

[7] I think this is debatable, but I won't challenge it here. For example, one might interpret Kimura's (1983) neutral theory of molecular evolution as demonstrating that just as there are instances of nonevolutionary changes in the phenotype structure of a population, there are also nonevolutionary changes in genotype space that make no difference to the lives of organisms and no difference to the course of evolution.

Modern Synthesis thinking implies a rigid distinction between those changes in population structure that are genuinely evolutionary and those that are not; it's all about genes. Situated Darwinism at least gives us a principled reason to think there should be no criterion of evolution. If evolution is the ecological phenomenon proposed by Situated Darwinism, that occurs as a consequence of the adaptive engagement of organisms with their affordances, we should expect that there is no such demarcation to be had.

With these Modern Synthesis dogmas swept away, we are free to approach the process of evolution anew. An updated account of evolution, one that is in line with Situated Darwinism should, at least, represent adaptive evolution not as the winnowing of passive form by an external environment, but as the interaction of agents and their affordances. It should take seriously the Darwinian insight that evolution is the direct consequence of what organisms do.

11.2   Organisms in evolution

I have been stressing that organisms are not mere objects of evolutionary forces. They are agents of evolutionary change. In pursuing their goals, in negotiating their affordance landscapes, in constructing their conditions of existence, organisms enact evolution. As such, they contribute to evolution in myriad ways. In fact, any process that contributes to the change (or stasis) in the origin or the frequency of intergenerationally stable forms is potentially an evolutionary process. Modern Synthesis evolutionary thinking typically excludes developmental plasticity, learning, cultural transmission, behaviour (Bateson and Gluckman 2011; Vane-Wright 2014) and ecological engineering from its roster of evolutionary processes. Situated Darwinism, on the other hand, counts many more kinds of processes as (potentially) genuinely evolutionary. It is here that the empirical advances of our current century really start to support an alternative picture. A brief if incomplete list may help to underscore organisms' active participation in their own evolution.

Buffering   The various activities of organisms can buffer the effects of mutations and environment on form. This occurs in a variety of ways. Behaviour, such as behavioural thermoregulation – diurnal and seasonal movements – render an organism less sensitive to environmental extremes (Bateson and Gluckman 2011). Metabolic systems such as thermoregulatory or immune systems diminish the threats raised by diseases, or excesses of temperature. Organisms manipulate their environments in ways that make them reliably propitious for survival and reproduction (Odling-Smee, Laland and Feldman 2003). In all these ways the activities of organisms help to structure the affordances on which form evolves. Developmental robustness buffers organisms against the potentially adverse effects of both genetic and extra-genetic perturbations. Phenotypic accommodation ameliorates the adverse consequences that may accrue to organisms due to changes in their form. Organisms accommodate to those changes, thereby reducing their potentially deleterious consequences.

Innovating   Organisms produce new phenotypes in response to the affordances they encounter. They initiate novel behaviours, novel forms in response to developmental challenges. They respond to environmental stresses by the regulation of genes, for example through novel DNA methylation (Dowen, Pelizzolaa and Schmitza 2012). But there are many more ways to innovate. Innovations can arise, for example, from the adaptive search of phenotype space by gene regulatory networks (Wagner 2102, 2014). Such novelties can be secured across generations through any number of means. If novel environmentally initiated changes are stable, or recurrent, they may contribute to the stability of new forms. Innovations can be made stable through learning and behaviour (Bateson 2014; Corning 2014), or cultural transmission, or through ecological engineering. Organisms construct and maintain their environments, which in turn alters the conditions under which their offspring develop (Turner 2000), thereby reconfiguring their affordances.

Phenotypic accommodation is another source of innovation. In accommodating to changes in the phenotype or environment, organisms produce further phenotypic novelties. Where the novelties in question are responses to recurrent, stable environmental changes, they may be especially strongly inheritable, without the need for any genetic changes:

Contrary to common belief, environmentally induced novelties may have greater evolutionary potential than do mutationally induced ones. They can be immediately recurrent in a population; are more likely than mutational novelties to correlate with particular environmental conditions and be subjected to consistent (directional) selection. (West Eberhard 2003: 498)

Many biologists are beginning to realise that the traditional distinctions between genetic and environmental innovations is a false one (Müller 2010). Every novelty is the result of co-ordinated response of the entire complex nonlinear developmental system: 'The realizing conditions for phenotypic evolution are embodied in developmental systems that are characterized by cellular self-organization, feedback regulations and environmental dependence' (Müller 2010: 322). West-Eberhard's position on the importance of phenotypic plasticity bears repeating here.

Responsive phenotype structure is the primary source of novel phenotypes. And it matters little from a developmental point of view whether the recurrent change we call a phenotypic novelty is induced by a mutation or by a factor in the environment. (West-Eberhard 2003: 503)

Müller (2010) and Newman and Müller (2000, 2007) further distinguish the role of genetic evolution from that of the epigenetic innovation. The function of genetic evolution, they contend, is not in the origin of innovations, but in making these stable: 'Genetic evolution, while facilitating innovation, serves a consolidating role rather than a generative one, capturing and routinizing morphological templates' (Müller 2010: 323).

Facilitating   Organisms facilitate adaptive evolution. The structure of gene regulatory networks, for example, not only buffers phenotypes against mutation, it acts as a capacitor for evolutionary change. Gene regulatory networks have the capacity to produce stable phenotypic outputs while absorbing and accumulating any number of mutations. This form of buffering allows a population to store up variations. In doing so it confers on populations the capacity to respond in an unpredictable variety of ways to changing genetic, epigenetic and environmental conditions.

The power of organismal robustness to promote evolution is vividly described by recent work from Andreas Wagner and colleagues (Wagner 2011, 2012). Wagner and his coworkers observe and model the dynamics of gene networks. This work was reviewed briefly in Chapter 7. It demonstrates three important ways in which gene regulatory networks facilitate evolution. They buffer development against genetic and environmental perturbations. They act as capacitors for evolutionary change, by storing up genetic variation that might be used in other contexts. They permit the efficient search of viable phenotypes, making the origin of new stable forms more likely (Wagner 2014).

But it is not just the internal architecture of gene networks that promotes adaptive evolution. Entire genomes are constructed in such a way that disposes populations of organisms to undergo adaptive evolution. Genomes comprise significant chunks of largely unchanging elements. For example, of the roughly 20,000 genes in humans, roughly 50 per cent are thought to be shared with fruit flies. These have undergone virtually no changes (with the exception of duplications) over the 780 million years or so since our most recent common ancestor.[8] In addition to these highly conserved elements, genomes also contain highly variable elements. The conserved core genomic elements underwrite a battery of 'core processes'. 'Conserved core processes represent the basic machinery of the multicellular organism . . .' (Müller 2010: 260). These core processes combine with highly variable genes across a range of epigenetic, cellular and organismal settings to produce an array of

---

[8] The date is from www.timetree.org/.

phenotypic variants. This arrangement of unchanging core processes coupled to highly variable, plastic features of development is labelled 'facilitated variation' (Kirschner and Gerhard 2005). The most important feature of facilitated variation is that it confers on organisms 'the capacity to generate phenotypic variation in response to genotypic variation' (Kirschner and Gerhard 2010: 262). In particular, the way in which facilitated variation produces phenotypic variants has three crucial implications for evolution:

(1) [I]t maximizes the amount of phenotypic variation for a given amount of genotypic variations; (2) it minimizes the lethality of phenotypic variation; (3) it produces phenotypic variation that is most appropriate to the environmental conditions, even conditions never before encountered in the lineage. (Kirschner and Gerhard 2010: 262)

Facilitated variation, like the structure of gene networks, confers on organismal development its characteristic robust adaptiveness, which in turn is what makes adaptive evolution adaptive.

One of the most important contributions of robustness is that of storing up genetic variation as a resource for future adaptive phenotypic responses. A vivid, but by no means unique, example is found in the heat shock protein. HSP 90 acts as a capacitor for evolution, by damping down phenotypic variation and allowing genetic variation to accumulate (Rohner et al. 2013). In normal circumstances, HSP 90 minimises the phenotypic variation that is due to standing genetic variation (Rutherford and Lindquist 1998). This genetic variation is normally 'silent', but with developmental changes, particularly changes to HSP 90 function, these variations can become expressed, leading to novel phenotypes.

Co-opting   Much of evolutionary novelty consists in the reuse of existing developmental resources in new contexts (Wagner 2000; Müller 2008). *Hox* genes in particular, remain unchanged across enormous stretches of time, yet homologous *Hox* genes can underwrite the development of widely divergent structures. The *lab* gene and *Hox 1.6* of mice are virtual base pair for base pair copies, yet *lab* is implicated in the development of arthropod mouthparts, for which there is no vertebrate homologue, while *Hox* 1.6 is instrumental in the development of the vertebrate hyoid apparatus, a structure that has no homologue in arthropods (DuBoule and Dollé 1989). Transplanting the vertebrate *Hox 4.2* gene into *Drosophila* causes the development of the same structures that the virtually identical arthropod homeobox gene *Dfd* causes (McGinnis et al. 1990). The horns of beetles appear to have originated from the co-option of genes normally associated with the growth of limbs (Shubin, Tabin and Carroll 2009). The crystallins that form the lens of the tetrapod eye are repurposed metabolic enzymes (Wagner 2014).

The general idea is that evolutionary novelties usually arise through the application of existing developmental structures and processes in new contexts (Müller 2008). Novelty is not, in general, the result of genetic mutation so much as the redeployment of existing resources in new developmental contexts.

... the evolution of organismal form is much less a direct consequence of mutational genetic innovation, as believed earlier, but rather depends on continuing shifts, recruitments and rewiring of regulatory interactions in development. Evolution seems to favour the generation of alternative genetic circuits which are subsequently co-opted-into new regulatory functions. (Müller 2008: 14)

Here again, the robustness and the repertoire of developmental systems are of crucial importance. The capacity of development to accommodate to new circumstances, new arrangements of developmental resources, seems to be the primary source of phenotypic novelty: 'accommodation involves the reuse of old pieces in new places' (West-Eberhard 2005a: 617).

The major conclusion about phenotypic variation that emerges from these studies is that when novelty is achieved in the course of variation and selection, the novel trait may contain rather little that is new. (Kirschner and Gerhard 2010: 258)

Stabilising   The intergenerational stability of form required for evolution can be achieved by any number of organismal processes. The varieties of inheritance channels underscores the range of ways in which the stability of form can be secured across generations (Jablonka and Avital 2006; Jablonka and Lamb 2010).

Soma-soma transmission, in which crucial developmental resources are passed from one generation to the next through the passage of symbionts is extremely common (Jablonka and Lamb 2008). An enormous variety of modes of soma-soma transmission exists. These include intrauterine transmission of food preferences (Bilko, Altbäcker and Hudson 1994). Grooming of rat young by their mother helps to secure stress resistance. Young that are groomed more frequently by their mothers, groom their own young more frequently (Weaver et al. 2004).

Intergenerational stability of form is also promoted through the maintenance of environmental features necessary for the proper development of offspring (Odling-Smee et al. 2003; Turner 2000). Niches are resources for proper development of offspring. Their construction and maintenance are means for securing the transgenerational stability of form. These are all ways in which the resemblance of parents to offspring can be secured, and there are myriad others.

The adaptive plasticity of development further underwrites the stability of form required for evolution. Thanks to plasticity, organisms can reliably produce their phenotypes across a wide range of unpredictable contexts. One way this can happen is highlighted by the work of Andreas Wagner (2011,

2012). In his connected neutral networks, a population can move from comprising novel, nonrobust regulatory networks to highly robust regulatory networks simply through the process of mutation. Random mutation can take a gene network on a 'random walk' through the neutral network, from an area in which networks have many nonviable neighbours, to areas in network space that have few nonviable neighbours. So random mutation, far from being simply an agent of random change, can introduce a positive inheritance bias. The structure of neutral networks illustrates how novel traits can be locked in or routinised through mutation (Müller and Newman 2005; Newman, Forgacs and Müller 2006).

Furthermore, even when organismal stability and inheritance are under-written by gene action, organismal plasticity contributes to the maintenance of the functional integrity of the genome. Cells actively cut, transpose, copy and fix their genomes. They do so in highly sensitive, adaptive ways.

A major assertion of many traditional thinkers about evolution ... is that living cells cannot make specific, adaptive use of their natural genetic engineering capacities. They make this assertion to protect their view of evolution as the product of random, undirected genome change. But their position is philosophical, not scientific, nor is it based on empirical observations. (Shapiro 2011: 55–56)

These considerations all point to the importance of adaptive plasticity – *purposiveness* – for the adaptive evolution of organismal form. Purposiveness consists in the capacity of organisms as agents to respond adaptively to genetic, epigenetic and environmental influences, in the production of viable, robust living things. This plasticity is evident at every level of biological organisation (Keller forthcoming). It is a prerequisite for adaptive evolution. The purposiveness that makes organisms agents appears to be a necessary precondition of adaptive evolution.

Conclusion

Evolution is an ecological phenomenon. It arises out of the purposive engagement of organisms with their conditions of existence. We might indeed consider this to be the lesson of cardinal importance to be drawn from *The Origin of Species*. The Modern Synthesis theory of evolution that has held sway over biology for almost a century has lost sight of this simple, but crucial perspective. Taking this lesson on board has enormous implications for the way we think of evolution. It suggests, for example, that the fractionation of evolutionary biology into discrete processes of inheritance, development, selection and mutation is a mistake. The component processes of evolution do not have discrete, proprietary causes. They are all jointly caused by the agency of organisms and their ecological relations with their affordances.

The more we understand that the processes of evolution are intimately intertwined and widely distributed, the less plausible it should seem to us that unit genes exert some kind of privileged control over the dynamics of evolution.

From this organismal perspective many of the orthodox Modern Synthesis distinctions fade into irrelevance. There is no distinction in kind between innate and acquired traits. There is no distinction in kind between novelties that arise from genes and those that arise from developmental or environmental sources. There is no criterion to distinguish between evolutionary and nonevolutionary events. An event is an evolutionary event only if it has some lasting impact on a lineage. Often this can only be ascertained in retrospect.

Situated Darwinism contends that the questions of most pressing importance for evolution revolve around the ways in which organisms constitute and hold in place the conditions for evolution. How do organisms regulate, structure and modify genome structure and function? How do organisms synthesise, regulate and marshal the resources of their genes and environments in development? How does the purposiveness of organism secure the pattern of resemblance and difference that constitutes inheritance? How do organisms transduce the inputs of their physical and social environments into intergenerationally stable structures and behaviours? These are questions for a new evolutionary biology. They begin with the recognition of organisms as natural, purposive entities, agents that enact evolution through the pursuit of their goals, in the 'struggle for life'.

References

Agrawal, A., C. Laforsch and R. Tollrian (1999). Transgenerational Induction of Defences in Animals and Plants. *Nature* 401: 60–63.

Alexander, S. (1920). *Space, Time, and Deity*, the Gifford Lectures for 1916–18, 2 vols. (London: Macmillan).

Allen, G.E. (2005). Mechanism, Vitalism and Organicism in Late Nineteenth and Twentieth-Century Biology: The Importance of Historical Context. *Studies in the History and Philosophy of Biology and the Biomedical Sciences* 36: 261–283.

Amundson, R. (1994). Two Concepts of Constraint: Adaptationism and the Challenge from Developmental Biology. *Philosophy of Science* 61: 556–578.

Amundson, R. (2001). Adaptation and Development: On the Lack of Common Ground. In S.H. Orzack and E. Sober (Eds.), *Adaptationism and Optimality* (pp. 303–334). Cambridge: Cambridge University Press.

Amundson, R. (2005). *The Changing Role of the Embryo in Evolution*. Cambridge: Cambridge University Press.

Anscombe, E. (1957). *Intention*. Cambridge: Cambridge University Press.

Aquinas, T. (2006). *Summa Theologica*. New Advent Organization. www.Newadvent.Org/Summa/1002.Htm.

Arber, A. (1964). *The Mind and the Eye*. Cambridge: Cambridge University Press.

Ariew, A. (1996). Innateness and Canalization. *Philosophy of Science* 63: S19–S27.

Ariew, A. (1999). Innateness Is Canalisation: A Defense of a Developmental Account of Innateness. In V. Hardcastle (Ed.), *Where Biology Meets Psychology: Philosophical Essays* (pp. 117–138). Cambridge, MA: MIT Press.

Ariew, A. (2008). Population Thinking. In M. Ruse (Ed.), *Oxford Handbook of Philosophy of Biology* (pp. 64–86). Oxford: Oxford University Press.

Ariew, A., and Z. Ernst. (2009). What Fitness Can't Be. *Erkenntnis* 71: 289–301.

Ariew, A., C. Rice and J. Rohwer (2015). Galtonian Explanations. *British Journal for the Philosophy of Science* 66: 635–658.

Aristotle (1995). *De Anima*. In T. Irwin and G. Fine (Eds.), *Aristotle Selections* (pp. 169–218). London: Hackett.

Aristotle (1996). *Physics*. Trans. R. Waterfield. Oxford: Oxford University Press.

Aristotle (N.D.). *On the Gait of Animals*. Trans. A.S.L. Farquharson. Raleigh, NC: Alex Catalogue.

Ayala, F. (1970). Teleological Explanations in Evolutionary Biology. *Philosophy of Science* 37(1): 1–15.

Baldwin, J.M. (1896). A New Factor in Evolution. *American Naturalist* 30: 441–451, 536–553.

Barandiaran, X., E. Di Paolo and M. Rohde (2009). Defining Agency: Individuality, Normativity, Asymmetry and Spatio-Temporality in Action (Vol. 1.0). *Journal of Adaptive Behavior*. M. Rohde and T. Ikegami (Eds.), Special Issue on Agency, pp. 1–13. http://Barandiaran.Net/Textos/Defining_Agency

Barnes, B., and J. Dupré (2008). *Genomes and What to Make of Them*. Chicago: Chicago University Press.

Bateson, P. (1976). Specificity and the Origins of Behavior. In J. Rosenblatt, R.A. Hinde, and C. Beer (Eds.), *Advances in the Study of Behavior*, Vol. 6 (pp. 1–20). New York: Academic Press.

Bateson, P. (2014). New Thinking about Biological Evolution. *Biological Journal of the Linnean Society* 112: 268–275.

Bateson, P., and P. Gluckman (2011). *Plasticity, Robustness, Development and Evolution*. Cambridge: Cambridge University Press.

Bateson, P., and M. Mameli (2007). The Innate and the Acquired: Useful Clusters or a Residual Distinction from Folk Biology? *Development and Psychology* 49: 818–831. DOI: 10.1002/Dev

Bechtel, W. (2006). *Discovering Cell Mechanisms: The Creation of Modern Cell Biology*. Cambridge: Cambridge University Press.

Bechtel, W., and A. Abrahamsen (2006). Explanation: A Mechanistic Alternative. *Studies in the History and Philosophy of the Biological and Biomedical Sciences* 36: 421–441.

Bechtel, W., and R.C. Richardson (1993). Emergent Phenomena and Complex Systems. In A. Beckermann, H. Flohr and J. Kim (Eds.), *Emergence or Reduction? Essays on the Prospects of Nonreductive Physicalism* (pp. 257–288). Berlin, New York: De Gruyter.

Bechtel, W., and R.C. Richardson (1998). Vitalism. In E. Craig (Ed.), *Routledge Encyclopedia of Philosophy*. London: Routledge. https://mechanism.ucsd.edu/teaching/philbio/vitalism.htm Mechanism.Ucsd.Edu/Teaching/Philbio/Vitalism.Htm

Becker, Jamie (2013). A Drop in the Ocean Is Teeming with Life. Woods Hole Institute Joint Program in Oceanographic/Applied Ocean Science and Engineering. Mit.Whoi.Edu/Student-Research?tid=1423&cid=118149

Bedau, M. (1991). Can Biological Teleology Be Naturalized? *Journal of Philosophy* 88: 647–655.

Bedau, M. (1996). *The Nature of Life*. In M. Boden (Ed.), *The Philosophy of Artificial Life* (pp. 257–288). Oxford: Oxford University Press.

Bedau, M. (1998). Where's the Good in Teleology? Reprinted in C. Allen, M. Bekoff, and G. Lauder (Eds.), *Nature's Purposes: Analyses of Function and Design in Biology* (pp. 261–291). Cambridge, MA: MIT Press.

Beisson, J. (2011). Preformed Cell Structure and Cell Heredity. *Prion* 2 (1): 1–8.

Bell, G. (2008). *Natural Selection: The Mechanism of Evolution* (2nd ed.). Oxford: Oxford University Press.

Bergman, A., and M. Siegal. (2002). Evolutionary Capacitance as a General Feature of Complex Gene Networks. *Nature* 424: 549–552.

Bianconi, E., A. Piovesan, F. Facchin, A. Beraudi, R. Casadei, F. Frabetti, L. Vitale, M. C. Pelleri, S. Tassani, F. Piva, S. Perez-Amodio, P. Strippoli and S. Canaider (2013). An Estimation of the Number of Cells in the Human Body. *Annals of Human Biology* 40 (6): 463–471.

Bilko, A., V. Altbäcker, and R. Hudson (1994). Transmission of Food Preferences in the Rabbit: The Means of Information Transfer. *Physiology and Behaviour* 56: 907–912.

Bird, A. (2000). *Kuhn*. London: Routledge.

Bird, A. (2007). *Nature's Metaphysics: Laws and Properties*. Oxford: Oxford University Press.

Bolker, J. (2000). Modularity in Development and Why It Matters to Evo-Devo. *American Zoologist* 40: 770–776.

Bonner, J.T. (1958). *The Evolution of Development: Three Special Lectures Given at University College, London*. Cambridge: Cambridge University Press.

Bonner, J.T. (1982). *Evolution and Development*. Berlin: Springer.

Bonner, J.T. (2010). Brainless Behavior: A Myxomycete Chooses a Balanced Diet. *Proceedings of the National Academy of Sciences* 107: 5267–5268. DOI: 10.1073/Pnas.1000861107

Boyle, M., and D. Lavin (2010). Goodness and Desire. In S. Tenenbaum (Ed.), *Desire, Practical Reason, and the Good* (pp. 202–233). Oxford: Oxford University Press.

Brandon, R. (1990). *Adaptation and Environment*. Princeton: Princeton University Press.

Brandon, R. (2014). Natural Selection. In E.N. Zalta (Ed.), *The Stanford Encyclopedia of Philosophy* (Spring 2014 Edition). Plato.Stanford.Edu/Archives/Spr2014/Entries/Natural-Selection/.

Branzei, D., and M. Foiani (2008). Regulation of DNA Repair throughout the Cell Cycle. *Nature Reviews (Molecular Cell Biology)* 9: 297–308.

Brigandt, I., and A. Love (2012). Reductionism in Biology. In E.N. Zalta (Ed.), *The Stanford Encyclopedia of Philosophy* (Summer 2012 Edition). Plato.Stanford.Edu/Archives/Sum2012/Entries/Reduction-Biolog.

Broad, C.D. (1925). *Mind and Its World*. London: Routledge and Kegan Paul.

Brown, R. (2013). What Evolvability Really Is. *British Journal for the Philosophy of Science*. 10.1093/Bjps/Axt014

Brunnander, B. (2007). What Is Selection? *Biology and Philosophy* 22: 231–246.

Bulgakov, M. (1986). *The Life of Monsieur De Molière: A Portrait by Mikhail Bulgakov*. Trans. Mirra Ginsburg. Toronto: Penguin Books, Canada.

Cain, J. (2009). Rethinking the Synthesis Period in Evolutionary Studies. *Journal of the History of Biology* 42: 621–648.

Calder, W.A. (1978). The Kiwi. *Scientific American* 239: 132–142.

Callebaut, W. (1993). *Taking the Naturalist Turn, or How Real Philosophy of Science Is Done*. Chicago: Chicago University Press.

Camazine, S., J.-L. Deneubourg, N.R. Franks, J. Sneyd, G. Theraulaz and E. Bonabeau (2001). *Self-Organization in Biological Systems*. Princeton: Princeton University Press.

Carroll, S.B., J.K. Grenier and S.D. Weatherbee (2000). *From DNA to Diversity: Molecular Genetics and the Evolution of Animal Design*. London: Wiley-Blackwell.

Cartwright, N. (1999). *The Dappled World: A Study in the Boundaries of Science*. Cambridge: Cambridge University Press.

Cartwright, N. (2004). Causation: One Word, Many Things. *Philosophy of Science* 71: 805–819.

Cartwright, N. (2010). Models: Parables vs. Fables. In T. Frigg and M. Hunter (Eds.), *Beyond Mimesis and Convention: Representation in Art and Science* (pp. 19–31). New York: Springer.

Chemero, A. (2003). An Outline of a Theory of Affordances. *Ecological Psychology 15*: 181–195.

Chen, X., J.R. Bracht, A.D. Goldman, E. Dolzhenko, D.M. Clay, E.C. Swart, D.H. Perlman, T.G. Doak, A. Stuart, C.T. Amemiya, R.P. Sebra and L.F. Landweber (2014). The Architecture of a Scrambled Genome Reveals Massive Levels of Genomic Rearrangement during Development. *Cell* 158: 1187–1198.

Christner, B.C., John C. Priscu, Amanda M. Achberger, Carlo Barbante, Sasha P. Carter, Knut Christianson, Alexander B. Michaud, Jill A. Mikucki, Andrew C. Mitchell, Mark L. Skidmore, Trista J. Vick-Majors and the WISSARD Science Team (2014). A Microbial Ecosystem beneath the West Antarctic Ice Sheet. *Nature* 512: 310–313. DOI: 10.1038/Nature13667

Churchill, F.B. (1974). William Johannsen and the Genotype Concept. *Journal of the History of Biology* 7: 5–30.

Ciccia, A., and S.J. Elledge (2010). The DNA Repair Response: Making It Safe to Play with Knives. *Molecular Cell* 40: 179–204.

Ciliberti, S., O.C. Martin and A. Wagner (2007a). Robustness Can Evolve Gradually in Complex Regulatory Gene Networks with Varying Topology. *PLoS Comput Biol* 3 (2): E15. DOI: 10.1371/Journal.Pcbi.0030015

Ciliberti, S., O.C. Martin and A. Wagner (2007b). Innovation and Robustness in Complex Regulatory Gene Networks. *PNAS* 104 (34): 13591–13596.

Clark, A., and D. Chalmers (1998). The Extended Mind. *Analysis* 58: 10–23.

Clayton, P., and P. Davies (2006). *The Re-Emergence of Emergence: The Emergentist Hypothesis from Science to Religion*. Oxford: Oxford University Press.

Collins, F.S. (1999). Medical and Societal Consequences of the Human Genome Project. *New England Journal of Medicine* 341: 28–37.

Conklin, E.G. (1921). Mechanism, Vitalism, and Teleology. *The Rice Institute Pamphlet*. 8: 351–380.

Corning P. (2014). Evolution 'On Purpose': How Behaviour Has Shaped the Evolutionary Process. *Biological Journal of the Linnean Society* 112: 242–260.

Craver, C. (2007). *Explaining the Brain: Mechanisms and the Mosaic Unity of Neuroscience*. Oxford: Oxford University Press.

Craver, C. (2013). Functions and Mechanism: A Perspectival View. In P. Huneman (Ed.), *Functions Selection and Mechanisms* (pp. 133–158). Dordrecht: Springer.

Craver, C., and L. Darden (2013). *In Search of Mechanisms: Discoveries across the Life Sciences*. Chicago: University of Chicago Press.

Crick, F.J. (1958). On Protein Synthesis. *Symposia of the Society for Experimental Biology* 12: 138–167.

Crick, F.J. (1967). *Of Molecules and Men* (pp. 24, 99). Seattle: University of Washington Press.

Crick, F.J. (1970). The Central Dogma of Molecular Biology. *Nature* 227: 561–563.

Curley, E. (Ed.) (1994). *Spinoza Reader*. Princeton, NJ: Princeton University Press.

Danchin, E., and A.S. Pocheville (2014). Inheritance Is Where Physiology Meets Evolution. *Journal of Physiology* 592(11): 2307–2317.

Danchin, E., A. Charmontier, F.A. Champagne, A. Mesoudi, B. Pujol and S. Blanchet (2011). Beyond DNA: Integrating Inclusive Inheritance into an Extended Theory of Evolution. *Nature Reviews Genetics* 12 (July): 475–486.

Darwin, C. (1859 [1968]). *The Origin of Species*. London: Penguin.

Davidson, D. (1963). Actions, Reasons and Causes. *Journal of Philosophy* 60: 685–700.

Davidson, D. (1980). *Agency*. In D. Davidson (Ed.), *Essays on Actions and Events* (pp. 43–62). Oxford: Oxford University Press.

Davidson, E.H. (2006). *The Regulatory Genome: Gene Regulatory Networks in Development and Evolution*. London: Academic Press.

Davidson, E.H. (2010). Emerging Properties of Animal Gene Regulatory Networks. *Nature* 468: 911–920.

Davidson, E.H., and D.H. Erwen (2006). Gene Regulatory Networks and the Evolution of Animal Body Plans. *Science* 311: 796–800.

Davidson, E.H., and M.S. Levine (2006). Properties of Gene Regulatory Networks. *PNAS* 105: 20063–20066.

Davies, P.C.W. (2012). The Epigenome and Top-Down Causation. *Interface Focus* 2: 42–48.

Dawkins, R. (1976). *The Selfish Gene*. Oxford: Oxford University Press.

Dawkins, R. (1982): *The Extended Phenotype*. Oxford: Oxford University Press.

Dawkins, R. (1986). *The Blind Watchmaker*. Oxford: Oxford University Press.

Dawkins, R. (1989). *The Selfish Gene* (2nd ed.). Oxford: Oxford University Press.

Dawkins, R. (1998). Universal Darwinism. Reprinted in D. Hull and M. Ruse (Eds.), *Oxford Readings in Philosophy of Biology* (pp. 15–37). Oxford: Oxford University Press.

Dawkins, R. (1999). *The Extended Phenotype: The Long Reach of the Gene* (2nd ed.). Oxford: Oxford University Press.

Dawkins, R. (2004). Extended Phenotype, But Not Too Extended – A Reply to Laland, Turner and Jablonka. *Biology and Philosophy* 19: 377–396.

Deacon, T.W. (2011). *Incomplete Nature: How Mind Emerged from Matter*. New York: W.W. Norton and Company.

De Bakker, M.A.G., D.A. Fowler, K. Den Oude, E.M. Dondorp, M.C.G. Navas, J.O. Horbanczuk, J.-Y. Sire, D. Szczerbinska and M.K. Richardson (2013). Digit Loss Interplay between Selection and Constraints. *Nature* 500: 445–448.

De Beer, G. (1938). Embryology and Evolution. In G. De Beer (Ed.), *Evolution: Essays On Aspects of Evolutionary Biology* (pp. 57–78). Oxford: Clarendon Press.

Debat, V., and P. David (2001). Mapping Phenotypes: Canalization, Plasticity and Developmental Stability. *Trends in Ecology and Evolution* 16 (10): 555–561.

Dennett, D. (1995). *Darwin's Dangerous Idea: Evolution and the Meanings of Life*. New York: Touchstone Books.

Depew, D. (2011). Adaptation as a Process: The Future of Darwinism and the Legacy of Theodosius Dobzhansky. *Studies in the History of Biology and the Biomedical Sciences* 42: 89–98.

Depew, D. (forthcoming). In D. Walsh and P. Huneman (Eds.), *Challenges to Evolutionary Theory*. Oxford: Oxford University Press.

Depew, D., and B. Weber (1995). *Darwinism Evolving: Systems Dynamics and the Geneaology of Natural Selection*. Cambridge, MA: MIT Press.

Depew, D., and B. Weber (2011). The Fate of Darwinism: Evolution after the Modern Synthesis. *Biological Theory* 6: 89–102.

Descartes, R. (1647 [1985]). *Principia Philosophiae*. Excerpted from *Selected Works*. Cottingham, J.T. Stoothoff and D. Murdoch (Eds.). Cambridge: Cambridge University Press.

Di Paolo, E. (2005). Autopoiesis, Adaptivity, Teleology, Agency. *Phenomenology and the Cognitive Sciences*. 4: 429–452.

Dickens, T.E., and Q. Rahman (2012). The Extended Evolutionary Synthesis and the Role of Soft inheritance in Evolution. *Proceedings of the Royal Society. Series B*. 279 (1740): 2913–2921. DOI: 10.1098/Rspb.2012.0273

Dobzhansky, T. (1937). *Genetics and the Origin of Species*. New York: Columbia University Press.

Dowen, R.H., M. Pelizzolaa, R.J. Schmitz, R.A., Lister, R., Dowen, J.M., Nery, J.R. and Ecker, J.R. (2012). Widespread Dynamic DNA Methylation in Response to Biotic Stress. *PNAS*. 109(32):E2183–91. DOI: 10.1073/pnas.1209329109

Dretske, F. (1986). Misrepresentation. In R. Bogdan (Ed.). *Belief: Form, Content, and Function* (pp. 17–36). Oxford: Oxford University Press.

Duboule, D., and P. Dollé (1989). The Structural and Functional Organization of the Murine *Hox* Family Resembles That of Drosophila Homeotic Genes. *Embo* 8: 1497–1505.

Dupré, J. (2012). *Processes of Life: Essays in the Philosophy of Biology*. Oxford: Oxford University Press.

Dupré, J. (2013). Living Causes. *Proceedings of the Aristotelian Society Supplementary Volume* 87: 19–38.

Earnshaw-Whyte, E. (2012). increasingly Radical Claims about Heredity and Fitness *Philosophy of Science* 79: 396–412.

Echten, S., and J. Borovitz (2013). Epigenomics: Methylation's Mark on Inheritance. *Nature* 495: 181–182. DOI:10.1038/Nature11960

Ellis, B. (2007). *Scientific Essentialism*. Cambridge: Cambridge University Press.

Ellis, G.F. (2012). *On the Nature of Causation in Complex Systems*. Unpublished Manuscript. Retrieved from www.Mth.Uct.Ac.Za/~Ellis/Top-Down_20Ellis.Pdf (accessed 16 January 2013).

Encyclopedia Britannica (2013). General Sherman Tree. http://www.britannica.com/pl ace/General-Sherman (accessed 3 September 2015).

Evans, E. (2013). Soil Life. North Carolina State University, College of Arts and Sciences Cooperative Extension, Horticultural Sciences. www.Ces.Ncsu.Edu/Depts/ Hort/Consumer/Quickref/Soil/Soillife.Html

Falk, R. (1986). What Is a Gene? *Studies in the History and Philosophy of Science* 17: 133–173.

Ferguson-Smith, A.C. (2011). Genomic Imprinting: The Emergence of an Epigenetic Paradigm. *Nature Review Genetics* 12: 565–575.

Fisher, R.A. (1918). The Correlation between Relatives on the Supposition of Mendelian Inheritance. *Transactions of the Royal Society of Edinburgh* 52: 399–433.

Fisher, R.A. (1930). *The Genetical Theory of Natural Selection*. Oxford: Clarendon Press.

Flyvbjerg, H., and B. Lautrup (1992). Evolution in a Rugged Fitness Landscape. *Physical Review* 46: 6714–6723.

Fodor, J. (1974). Special Sciences (Or: the Disunity of Science as a Working Hypothesis). *Synthese* 28: 97–115.

Fodor, J. (1987). *Psychosemantics: The Problem of Meaning in Philosophy of Mind.* Cambridge, MA: MIT Press.

Forgacs, G., and S. Newman (2005). *Biological Physics of the Developing Embryo.* Cambridge: Cambridge University Press.

Fulda, F. (ms). Between Mechanisms and Intellectualism: The Case of Bacterial Cognition. *Unpublished manuscript.*

Fulda, F. (2015). *An Ecological Approach to Agency.* PhD Dissertation, University of Toronto.

Galton, D. (2009). Did Darwin Read Mendel? *Quarterly Journal of Medicine* 209: 587–589.

Ganeri, J. (2011). Emergentisms, Ancient and Modern. *Mind* 120: 671–703.

Garber, D. (1998 [2003]). Descartes, René. In E. Craig (Ed.), *Routledge Encyclopedia of Philosophy.* London: Routledge. Retrieved from http://www.Rep.Routledge.Com/A rticle/DA026SECT3

Garber, D. (2013). Remarks on the Pre-History of the Mechanical Philosophy. In D. Garber and S. Roux (Eds.), *The Mechanization of Natural Philosophy* (pp. 3–26). Boston: Springer.

Garfinkel, A. (1981). *Forms of Explanation: Rethinking Questions in Social Theory.* New Haven: Yale University Press.

Garrett, B. (2010). Vitalism versus Emergent Materialism. In S. Normandin and C.T. Wolfe (Eds.), *Vitalism and the Scientific Image in Post-Enlightenment Life Science, 1800–2010, History, Philosophy and Theory of the Life Sciences* 2 (pp. 127–154). Dordrecht: Springer. DOI: 10.1007/978–94-007–2445-7_6

Garstang, W. (1922). The Theory of Recapitulation: A Critical Re-Statement of the Biogenetic Law. *Linnean Journal of Zoology* 35: 81–101.

Gayon, J. (2000). From Measurement to Organization: A Philosophical Scheme for the History of the Concept of Inheritance. In P.J. Beurteon, R. Falk and H.-J. Rheinberge (Eds.), *The Concept of the Gene in Development and Evolution: Historical and Epistemological Perspectives* (pp. 60–90). Cambridge: Cambridge University Press.

Gerhard, J.C., and M. Kirschner (2005). *The Plausibility of Life: Resolving Darwin's Dilemma.* New York: Norton.

Gerhard, J.M., and M. Kirschner (2007). The Theory of Facilitated Variation. *PNAS* 104: 8582–8589.

Ghiselin, M. (1993). Darwin's Language May Have Been Teleological, but His Thinking Was a Different Matter. *Biology and Philosophy* 9: 482–492.

Gibson, G. (2002). Getting Robust about Robustness. *Current Biology* 12: R347–R349.

Gibson, G., and G. Wagner (2000). Canalisation in Evolutionary Genetics: A Stabilizing theory? *Bioessays* 22: 372–380.

Gibson, J.J. (1979). *The Ecological Approach to Visual Perception.* Boston: Houghton Mifflin.

Gibson, K. (1994). General Introduction: Animal Minds, Human Minds. In K. Gibson and T.I. Parker (Eds.), *Tools, Language and Cognition in Human Evolution* (pp. 3–19). Cambridge University Press.

Gilbert, S. (1999). The Role of Embryonic Induction in Creating Self. In F. Tauber (Ed.), *Organisms and the Origin of Self. Boston Studies in Philosophy of Science* 129: 341–360. Boston: Kluwer.

Gilbert, S. (2003a). The Morphogenesis of Evolutionary Developmental Biology. *International Journal of Developmental Biology* 47: 467–477.

Gilbert, S. (2003b). The Genome in Its Ecological Setting. *Annals of the New York Academy of Science* 981: 202–218.

Gilbert, S., and D. Epel (2009). *Ecological Developmental Biology.* Sunderland, MA: Sinauer.

Gillespie, J. (1977). Natural Selection for Variances in Offspring Number: A New Evolutionary Principle. *American Naturalist* 111: 1010–1014.

Ginsborg, H. (2001). Kant on Understanding Organisms as Natural Purposes. In E. Watkins, *Kant and the Sciences* (pp. 231–258). Oxford: Oxford University Press.

Glennan, S. (1996). Mechanisms and the Nature of Causation. *Erkenntnis* 44: 49–71.

Glennan, S. (2002). Rethinking Mechanistic Explanation. *Philosophy of Science* 64: 605–626.

Godfrey-Smith, P. (1996). *Complexity and the Function of Mind in Nature.* Cambridge: Cambridge University Press.

Godfrey-Smith, P. (2000). On the Theoretical Role of 'Genetic Coding'. *Philosophy of Science* 67: 26–44.

Godfrey-Smith, P. (2001a). Organism, Environment and Dialectics. In R. Singh, C. Krimbas, D. Paul and J. Beatty (Eds.), *Thinking about Evolution.* Cambridge: Cambridge University Press.

Godfrey-Smith, P. (2001b). Three Kinds of Adaptationism. In S.H. Orzack and E. Sober (Eds.), *Adaptationism and Optimality* (pp. 335–357). New York: Cambridge University Press.

Godfrey-Smith, P. (2009). *Darwinian Populations and Natural Selection.* Oxford: Oxford University Press.

Goodwin, B. (1994). *How the Leopard Changed Its Spots.* Princeton: Princeton University Press.

Gould, S.J. (1977). *Ontogeny and Phylogeny.* Cambridge, MA: Harvard University Press.

Gould, S.J. (2002). *The Structure of Evolutionary Theory.* Cambridge, MA: Harvard, Belknap Press.

Gould, S.J., and R.C. Lewontin (1979). The Spandrels of San Marco and the Panglossian Paradigm: A Critique of the Adaptationist Programme. *Proceedings of the Royal Society London, Series B,* 205: 581–598.

Greenspan, R.J. (2001). The Flexible Genome. *Nature Reviews Genetics,* 2: 383–387.

Grene, M. (1961), Statistics and Selection. *British Journal for the Philosophy of Science* 12: 25–42.

Grene, M. (1974). Aristotle and Modern Biology. Reprinted in M. Greene, *Understanding Nature: Essay in the Philosophy of Biology.* Boston Studies in the Philosophy of Science, Vol. XXIII, R.D.S. Cohen and M.W. Wartofsky (Eds.), pp. 74–107.

Griesemer, J.R., and W.C. Wimsatt (1989). Picturing Weismannism: A Case Study of Conceptual Evolution. In M. Ruse (Ed.), *What Philosophy of Biology Is. Essays Dedicated to David Hull* (pp. 75–137). Dordrecht, Netherlands: Kluwer Academic Publishers.

Griffiths, P.E. (2001). Genetic information: A Metaphor in Search of a theory. *Philosophy of Science* 68: 394–412.

Griffiths, P.E., and R.D. Gray (1994). Devepmental Systems and Evolutionary Explanation. *The Journal of Philosophy* 91: 277–304.

Griffiths, P.E., and R.D. Gray (2001). Darwinism and Developmental Systems. In S. Oyama, P.E. Griffiths and R.D. Gray (Eds.), *Cycles of Contingency: Developmental Systems and Evolution* (pp. 195–218). Cambridge, MA: MIT Press.

Griffiths, P.E., and R. Gray (2005). Discussion: Three Ways to Misunderstand Developmental Systems Theory. *Biology and Philosophy* 20: 417–25. DOI: 10.1007/S10539-004-0758-1

Griffiths, P.E., and K. Stotz (2007). Gene. In M. Ruse (Ed.) (2005), *Cambridge Companion to the Philosophy of Biology* (pp. 85–103). Cambridge: Cambridge University Press.

Griffiths, P.E., and K. Stotz (2013). *Genetics and Philosophy: An Introduction.* Cambridge: Cambridge University Press.

Griffiths, P.E., and J. Tabery (2014). Developmental Systems Theory: What Does It Explain, and How Does It Explain It? *Advances in Child Development and Behavior* 44: 65–94.

Guerrero-Bosagna C., M. Savenkova, M.M. Haque, E. Nilsson and M.K. Skinner (2013). Environmentally Induced Epigenetic Transgenerational inheritance of Altered Sertoli Cell Transcriptome and Epigenome: Molecular Etiology of Male infertility. *Plos ONE* 8 (3): E59922. DOI: 10.1371/Journal.Pone.0059922

Guyer, P. (2005). *Organisms and the Unity of Science.* In P. Guyer (Ed.), Kant's System of Nature and Freedom: Selected Essays (pp. 86–111). Oxford: Oxford University Press.

Haig, D. (2011). Lamarck Ascending! A Review of *Transformations of Lamarckism: From Subtle Fluids to Molecular Biology*, S.B. Gissis and E. Jablonka (Eds.). Cambridge, MA: MIT Press, *Philosophy and Theory in Biology* 3: E204.

Haldane, J.B.S. (2008). A Defence of Beanbag Genetics. *International Journal of Epidemiology* 37 (3): 435–442. DOI: 10.1093/Ije/Dyn056

Hall, B.K. (1999). *Evolutionary Developmental Biology.* Amsterdam: Kluwer.

Hall, B.K. (2012). Evolutionary Developmental Biology (Evo-Devo): Past, Present, and Future. *Evolution: Education and Outreach* 5: 184–193.

Hallgrímsson, B., K. Willmore and B.K. Hall (2002). Canalization, Developmental Stability, and Morphological integration in Primate Limbs. *Yearbook of Physical Anthropology* 45: 131–158.

Hamburger, V. (1980). Embryology and the Modern Synthesis in Evolutionary Biology. In E. Mayr and W. Provine (Eds.), *The Evolutionary Synthesis* (pp. 97–112). Cambridge, MA: Harvard University Press.

Hamburger, V., H.G.E. Allen and J. Maienschein (1999). Hans Spemann on Vitalism in Biology: Translation of a Portion of Spemann's *Autobiography*. *Journal of the History of Biology* 32: 231–243.

Hamel, A., C. Fisch, L. Combettes, P. Dupuis-Williams and C.N. Baroud (2011). Transitions between Three Gaits in *Paramecium* Scape. *Proceedings of the National Academy of Sciences* 108 (18): 7290–7295.

Hankinson, J. (1998). *Cause and Explanation in Ancient Greek Thought.* Oxford: Oxford University Press.

Harré, R. (1986). *Varieties of Realism: A Rationale for the Natural Sciences.* Oxford: Basil Blackwell.

Haugeland, J. (1998). *Having Thought: Essays in the Metaphysics of Mind*. Chicago: University of Chicago Press.

Heijmans, Bastiaan T., Elmar W. Tobi, Aryeh D. Stein, Hein Putter, Gerard J. Blauw, Ezra S. Susser, P. Eline Slagboom and L.H. Lumey (2008). Persistent Epigenetic Differences Associated with Prenatal Exposure to Famine in Humans. *PNAS* 105:17046–17049; Published Ahead of Print 27 October, 2008. DOI: 10.1073/Pnas.0806560105

Heil, M., S. Greiner, H. Meimberg, R. Krüger, J-L. Noyer, G. Heubl, K.E. Linsenmair and W. Boland (2004). Evolutionary Change from Induced to Constitutive Expression of an Indirect Plant Resistance. *Nature* 430: 205–208. DOI: 10.1038/Nature02703

Helenterä, H., and Üller, T. (2010). The Price Equation and Extended Inheritance. *Philosophy and Theory in Biology* 2: E101.

Hendrikse, J.L., T.E. Parsons and B. Hallgrímsson (2007). 'Evolvability as the Proper Focus of Evolutionary Developmental Biology. *Evolution and Development* 9: 393–401.

Henning, B.G., and A.C. Scarfe (2013). *Beyond Mechanism: Putting Life Back into Biology*. Toronto: Lexington Books.

Henry, D. (2013). Optimality and Teleology in Aristotle's Natural Science. *Oxford Studies in Ancient Philosophy*, Vol. 45, pp. 225–264.

Hernández-Hernández, V., K.J. Niklas, S.A. Newman and M. Benítez (2012). Dynamical Patterning Modules in Plant Development and Evolution. *International Journal of Developmental Biology* 2012: 56(9): 661–674. DOI: 10.1387/Ijdb.120027mb

Hodge, M.J. S. (1987). Natural Selection as a Causal, Empirical, and Probabilistic theory. In L. Kruger, G. Gigenrenzer and M.S. Morgan (Eds.), *The Probabilistic Revolution*, Vol. 2 (pp. 233–270). Cambridge, MA: MIT Press.

Hornsby, J. (1997). *Simple Mindedness: A Defense of Naive Naturalism in the Philosophy of Mind*. Cambridge, MA: Harvard University Press.

Howell, E. (2014). How Many Stars in the Milky Way? *Space.Com* (21 May). www.Space.Com/25959-How-Many-Stars-Are-in-the-Milky-Way.Html

Hull, D. (1969). What Philosophy of Biology Is Not. *Journal of the History of Biology* 2: 241–268.

Hull, D. (1973). *Philosophy of Biological Science*. Englewood Cliffs, NJ: Prentice-Hall.

Huneman, P. (2010). Assessing the Prospects for a Return of Organisms in Evolutionary Biology. *History and Philosophy of the Life Sciences* 32: 341–372.

Huneman, P. (2014). A Pluralist Framework to Address Challenges to the Modern Synthesis in Evolutionary Theory. *Biological Theory* (9): 163–177. DOI: 10.1007/s13752-014-0174-y

Huxley, J. (1942). *Evolution: The Modern Synthesis*. London: Allen Unwin.

Ingold, T. (1986). *Culture and the Perception of the Environment*. Cambridge: Cambridge University Press.

Ingold, T. (1989). An Anthropologist Looks at Biology. *Man* (N.S.) 25: 208–229.

Irschick, D.J., R.C. Albertson, P. Brennnan, et al. (2013). Evo-Devo beyond Morphology: From Genes to Resource Use. *Trends in Ecology and Evolution* 28: 509–16.

Jablonka, E. (2001). The Systems of Inheritance. In Oyama, S., P.E. Griffiths and R. Gray (Eds.), *Cycles of Contingency* (pp. 99–116). Cambridge, MA: MIT Press.

Jablonka, E., and E. Avital (2006). Animal Innovation: The Origins and Effects of New Learned Behaviours. *Biology and Philosophy* 21: 135–141.

Jablonka, E., and S. Gissis (2011). *Transformation of Lamarckism*. Cambridge, MA: MIT Press.

Jablonka, E., and M. Lamb (1995). *Epigenetic Inheritance and Evolution: The Lamarckian Dimension*. Oxford: Oxford University Press.

Jablonka, E., and M. Lamb (2002). The Changing Concept of Epigenetics. *Annals of the New York Academy of Sciences* 981: 82–96.

Jablonka, E., and M. Lamb (2004). *Evolution in Four Dimensions: Genetic, Epigenetic, Behavioral, and Symbolic Variation in the History of Life*. Cambridge, MA: Bradford Books.

Jablonka, E., and M. Lamb (2008). Soft Inheritance: Challenging the Modern Synthesis Genetics. *Molecular Biology* 31: 289–395.

Jablonka, E., and M. Lamb (2010). Transgenerational Epigenetic Inheritance. In M. Pigliucci and G. Müller (Eds.), *Evolution: The Extended Synthesis* (pp. 137–174). Cambridge, MA: MIT Press.

Jablonka, E., and G. Raz (2009). Transgenerational Epigenetic Inheritance: Prevalence, Mechanisms, and Implications for the Study of Heredity and Evolution. *The Quarterly Review of Biology* 84 (2): 131–176.

Jackson, S.P., and J. Bartek. (2009). The DNA-Damage Response in Human Biology and Disease. *Nature* 461: 1071–1078.

Jacob, F, (1973). *The Logic of Life: A History of Heredity*. Trans. B.R. Spillman. New York: Pantheon.

Jacob, F., and J. Monod (1961). Genetic Regulatory Mechanisms in the Synthesis of Proteins. *Journal of Molecular Biology* 3: 318–356.

Jenkin, F. (1867). The Origin of Species. *North British Review*. June. www.Victorian web.Org/Science/Science_Texts/Jenkins.Html

Johannsen W. (1911). The Genotype Conception of Heredity. *American Naturalist* 45: 129–159.

Johansen, T. (2004). *Plato's Naturalism*. Oxford: Oxford University Press.

Johnson, M.R. (2005). *Aristotle on Teleology*. Oxford: Oxford University Press.

Jonas, H. (1966). *The Phenomenon of Life: Toward a Philosophical Biology*. Evanston: Northwestern University Press.

Juarrero, A. (2012). *Complex Dynamical Systems Theory*. Http://Cognitive-Edge.Com/ Uploads/Articles/100608%20Complex_Dynamical_Systems_theory.Pdf (accessed 16 January 2013).

Kant, I. (2000). *Critique of the Power of Judgment*. Trans. P. Guyer and E. Matthew. Cambridge: Cambridge University Press. (First published 1790)

Kaplan, J. (2013). Relevant Similarity and the Causes of Biological Evolution: Selection, Fitness and the Causes of Biological Evolution. *Biology and Philosophy* 28: 405–421.

Kauffman, S. (1993). *The Origins of Order: Self-Organization and Selection in Evolution*. Oxford: Oxford University Press.

Kauffman, S. (1995). *At Home in the Universe*. Oxford: Oxford University Press.

Kauffman, S. (2010). What Is Life? Was Schrödinger Right? In M. Bedau and C. Cleland (Eds.), *The Nature of Life: Classical and Contemporary Perspectives from Philosophy* (pp. 374–391). Cambridge: Cambridge University Press.

Kay, L. (2000). *Who Wrote the Book of Life?* Stanford: Stanford University Press.

Kazmierczak, J., and S. Kempe (2004). Calcium Build-Up in the Precambrian Sea: A Major Promoter in the Evolution of Eukaryotic Life. In J. Seckbach (Ed.), *Origins: Genesis, Evolution and Diversity of Life* (pp. 329–345). Dordrecht: Kluwer.

Keller, E.F. (2000). *The Century of the Gene.* Cambridge, MA: Harvard University Press.

Keller, E.F. (2011a). Genes, Genomes, Genomics *Biological Theory* 6: 132–140.

Keller, E.F. (2011b). *The Mirage of a Space between Nature and Nurture.* Raleigh: Duke University Press.

Keller, E.F. (2013). Genes as Difference Makers. In S. Krimsky and J. Gruber (Eds.), *Genetic Explanations: Sense and Nonsense* (pp. 329–345). Cambridge, MA: Harvard University Press.

Keller, E.F. (forthcoming). *The Genome.* Forthcoming in S. Richardson and H. Stevens, *The Age.* Raleigh: Duke University Press.

Kerr, B., and P. Godfrey-Smith. (2009). Generalization of the Price Equation for Evolutionary Change. *Evolution* 63: 531–536.

Kim, J. (1989). Mechanism, Purpose and Explanatory Exclusion. In J. Tomberlin (Ed.), *Philosophical Perspectives 3: Philosophy of Mind and Action Theory* (pp. 77–108). Atascadero, CA: Ridgeview.

Kim, J. (1999). Making Sense of Emergence. *Philosophical Studies* 95: 3–36.

Kim, J. (2006). Emergence: Core Ideas and Issues. *Synthese* 151: 547–559.

Kimura, M. (1983). *The Neutral Theory of Molecular Evolution.* Cambridge: Cambridge University Press.

King, N. (2004). The Unicellular Ancestry of Animal Development. *Developmental Cell* 7: 313–325.

King, N., M.J. Westbrook, S.L. Young, et al. (2008). The Genome of the Choanoflagellate *Monosiga Brevicollis* and the Origin of Metazoans. *Nature* 451: 783–788.

Kirschner, M., and J. Gerhard (2005). *The Plausibility of Life: Resolving Darwin's Dilemma.* New Haven: Yale University Press.

Kirschner, M., and J.C. Gerhard (1998). Evolvability. *Proceedings of the National Academy of Sciences USA* 95: 8420–8427.

Kirschner, M., and J.C. Gerhard (2010). Facilitated Variation. In M. Pigliucci and G.B. Muller (Eds.), *The Extended Synthesis* (pp. 253–280). Cambridge, MA: MIT Press.

Kitano, H. (2004). Biological Robustness. *Nature Reviews Genetics* 5: 826–837.

Kripke, S. (1972). *Naming and Necessity.* Oxford: Oxford University Press.

Kuhn, T. (1970). *The Structure of Scientific Revolutions* (2nd ed.). Chicago: University of Chicago Press.

Laland, K., T. Uller, M. Feldman, L. Sterelny, G.B. Müller, A. Moczek, E. Jablonka and J. Odling-Smee (2014). Does Evolutionary Theory Need a Rethink? Yes: Urgently. *Nature* 514: 161–164.

Laland, K., F.J. Odling-Smee and M.W. Feldman (2001). Niche Construction, Ecological Inheritance, and Cycles of Contingency in Evolution. In S. Oyama,

P. Griffiths and R. Gray (Eds.), *Cycles of Contingency: Developmental System and Evolution* (pp. 117–126). Cambridge, MA: MIT Press.

Lamm, E. (2011). The Metastable Genome. In E. Jablonka & S. Gissis (Eds.), *Transformations of Lamarckism: From Subtle Fluids to Molecular Biology* (pp. 345–355). Cambridge, MA: MIT Press.

Lamm, E. (2014). Inheritance Systems. In E.N. Zalta (Ed.), *The Stanford Encyclopedia of Philosophy* (Spring 2012 Edition). Plato.Stanford.Edu/Archives/Spr2012/Entries/inheritance-Systems

Lande, R. (1976). Natural Selection and Random Genetic Drift in Phenotypic Evolution. *Evolution* 30: 315–334.

Lange, M. (2013). Really Statistical Explanations and Genetic Drift. *Philosophy of Science*. 80: 169–188.

Laubichler, M. (2009). Evo-Devo: Historical and Conceptual Reflections. In M. Laublichler and J. Maienschein (Eds.), *Form and Function in Developmental Evolution* (pp. 10–46). Cambridge: Cambridge University Press.

Laubichler, M. (2010). Evolutionary Developmental Biology Offers a Significant Challenge to the Neo-Darwinian Paradigm. In F. Ayala and R. Arp (Eds.), *Contemporary Debates in Philosophy of Biology* (pp. 199–212). Malden, MA: John Wiley & Sons.

Lennox, J. (2001). Material and Formal Natures in Aristotle's *de Partibus Animalium*. In *Aristotle's Philosophy of Biology* (pp. 182–204). Cambridge: Cambridge University Press.

Lennox, J. (2009). Form, Essence and Explanation in Aristotle's Biology. In G. Anagnostopoulos (Ed.), *A Companion to Aristotle* (pp. 348–367). London: Blackwell.

Lennox, J. (2010). Βιος, Πραχισ, and The Unity of Life. In S. Follinger (Ed.), *Was ist 'Leben'?* (pp. 239–256). Stuttgart: Steiner Verlag.

Leunissen, M. (2010). *Explanation and Teleology in Aristotle's Science of Nature*. Cambridge: Cambridge University Press.

Levins, R. (1966). The Strategy of Model Building in Population Biology. Reprinted in E. Sober (Ed.) (1984), *Conceptual Issues in Revolutionary Biology* (pp. 18–27). Cambridge, MA: MIT Press.

Levins, R., and R.C. Lewontin (1985). *The Dialectical Biologist*. Cambridge, MA: Harvard University Press.

Lewens, T. (2004). *Organisms and Artifacts: Design in Nature and Elsewhere*. Cambridge, MA: MIT Press.

Lewens, T. (2009a). What's Wrong with Typological Thinking? *Philosophy of Science* 76: 355–371.

Lewens, T. (2009b). Seven Kinds of Adaptationism. *Biology and Philosophy* 24: 161–182.

Lewens, T. (2010). The Natures of Selection. *British Journal for the Philosophy of Science*. 61: 303–333.

Lewis, D.K. (1969). *Convention: A Philosophical Study*. Cambridge, MA: Harvard University Press.

Lewontin, R.C. (1974). *The Genetic Basis of Evolutionary Change*. New York: Columbia University Press.

Lewontin, R.C. (1978). Adaptation. *Scientific American* 239: 212–230.

Lewontin, R.C. (1983). Gene, Organism, and Environment. In D.S. Bendall (Ed.), *Evolution: from Molecules to Men* (pp. 273–285). Cambridge: Cambridge University Press.

Lewontin, R.C. (1985). The Organism as Subject and Object of Evolution. Reprinted in: R. Levins and R. Lewontin (1985), *The Dialectical Biologist*. Cambridge, MA: Harvard University Press Scientia 118: 85–106.

Lewontin, R.C. (2001a). *The Triple Helix: Genes, Organisms and Environments*. Oxford: Oxford University Press.

Lewontin, R.C. (2001b). The Problems of Population Genetics. In R. Singh, C. Krimbas, D.S. Paul and J. Beatty (Eds.), *Thinking about Evolution*, Vol. 1 (pp. 4–22). Cambridge: Cambridge University Press.

Lewontin, R.C. (2001c). Genes, Organisms and Environments. In S. Oyama, E. P. Griffiths and R. Gray, *Cycles of Contingency: Developmental Systems and Evolution* (pp. 59–66). Cambridge, MA: MIT Press.

Lipton, P. (2004). *Inference to the Best Explanation* (2nd ed.). London: Routledge.

López-Beltrán, C. (1994). Forging Heredity: From Metaphor to Cause, a Reification Story. *Studies in the History and Philosophy of Science* 25: 211–235.

Love, A. (2003). Evolutionary Morphology, Innovation, and the Synthesis of Evolutionary and Developmental Biology. *Biology and Philosophy* 18: 309–345.

Love, A. (2015). Conceptual Change and Evolutionary Developmental Biology. In A. Love (Ed.), *Conceptual Change in Biology: Scientific and Philosophical Perspectives on Evolution and Development* (pp. 1–54). Dordrecht: Springer.

Lumey, L.H. (1992). Decreased Birthweights in Infants after Maternal in Utero Exposure to the Dutch Famine of 1944–1945. *Pediatric and Perinatal Epidemiology*. 6: 240–253.

Lyon, P. (2005). The Biogenic Approach to Cognition. *Cognitive Processes*. DOI: 10.1007/S10339-005-0016-8.

Machamer, P., L. Darden and C. Craver (2000). Thinking about Mechanisms. *Philosophy of Science*. 57: 1–25.

Maienschein, J., and M. Laubichler (2014). Explaining Development and Evolution on the Tangled Bank. In R.P. Thompson and D.M. Walsh (Eds.), *Evolutionary Biology: Conceptual, Ethical, and Religious Issues* (pp. 151–171). Cambridge: Cambridge University Press.

Malthus, T. (1798). *An Essay on the Principle of Populations*. London: J. Johnson (W. W. Norton Edition, 1976, edited by Peter Appleman).

Mameli, M. (2004). Non-Genetic Selection and Non-Genetic Inheritance. *British Journal For the Philosophy of Science* 55: 35–71.

Mameli, M. (2005). The Inheritance of Features. *Biology and Philosophy* 20: 365–399.

Marzke, M. (1997). Precision Grip, Hand Morphology, and Tools. *American Journal of Physical Anthropology* 102: 91–110.

Matthen, M. (2009). Drift and 'Statistically Abstractive Explanation'. *Philosophy of Science* 76: 464–487.

Matthen, M., and A. Ariew. (2002). Two Ways of Thinking about Fitness and Selection. *Journal of Philosophy* 99: 58–83.

Matthen, M., and A. Ariew (2009). Selection and Causation. *Philosophy of Science* 76: 201–223.

Matthews, G. (2002). *The Philosophy of Merleau-Ponty*. London: Routledge.

Maynard Smith, J. (1969). The Status of Neo-Darwinism. In C.H. Waddington (Ed.), *Toward a Theoretical Biology*, pp. 82–89. Edinburgh: Edinburgh University Press.

Maynard Smith, J. (1982). *Evolution and the Theory of Games*. Cambridge: Cambridge University Press.

Maynard Smith, J. (1989). *Evolutionary Genetics*. Oxford: Oxford University Press.

Maynard Smith, J. (1998). *Evolutionary Genetics*, 2nd ed. Oxford: Oxford University Press.

Maynard Smith, J. (2000). The Concept of Information in Biology. *Philosophy of Science* 67: 177–194.

Maynard Smith, J., R. Burian, S. Kauffman, P. Alberch, J. Campbell, B. Goodwin, R. Lande, Da. Raup and L. Wolpert (1985). Developmental Constraints and Evolution. *Quarterly Review of Biology* 60: 265–287.

Mayr, E. (1963). *Animal Species and Evolution*. Cambridge, MA: Harvard University Press.

Mayr, E. (1975). *Evolution and the Diversity of Life*. Cambridge, MA: Harvard University Press

Mayr, E. (1976). *Towards a New Philosophy of Biology*. Cambridge, MA: Harvard University Press.

Mayr, E. (1982). *The Growth of Biological Thought*. Cambridge, MA: Harvard University Press.

Mayr, E. (1988). The Multiple Meanings of Teleological. In *Towards a New Philosophy of Biology: Observations of an Evolutionist* (pp. 38–66). Cambridge, MA: Harvard University Press.

Mayr, R. (1997). The Objects of Selection. *Proceedings of the National Academy of the Sciences* 91: 2091–2094.

Mayr, E., and W. B. Provine (Eds.) (1980). *The Evolutionary Synthesis: Perspectives on the Unification of Biology*. Cambridge, MA: Harvard University Press.

McClintock, B. (1984). The Significance of Responses of the Genome to Challenge. *Science* 226: 792–801.

McCosh, R.B., H. Chen, N.J. Schork and J.R. Ecker (2013). Patterns of Population Epigenomic Diversity. *Nature* 495: 193–200. DOI: 10.1038/Nature11968.

McGinnis, N., M.A. Kuziora and W. Mcginnis (1990). Human Hox and Drosophila Deformed Encode Similar Regulatory Specificities in Drosophila Embryos and Larvae. *Cell* 63: 969–976.

McLaughlin, P. (2000). *What Function Explains*. Cambridge: Cambridge University Press.

McLaughlin, P. (2014). The Impact of Newton on Biology on the Continent in the Eighteenth Century. In S. Mandelbrote and H. Pulte (Eds.), *The Reception of Isaac Newton in Europe* (pp. 1–23). London: Bloomsbury Academic. www.Philosophie.Uni-Hd.De/Md/Philsem/Personal/Mclaughlin_Newton_Biology.Pdf

Meir, E.G., G. Von Dassow, E. Munro and G. Odell (2002). Robustness, Flexibility, and the Role of Lateral inhibition in the Neurogenic Network. *Current Biology* 12: 778–786.

Mendel, G. (1866). Versuche Über Plflanzenhybriden. *Verhand- Lungen Des Naturforschenden Vereines in Brünn, Bd. IV Für Das Jahr 1865*, Abhandlungen, 3–47 (Bateson Translation (1901)).

Merlin, F. (2010). Evolutionary Chance Mutation: A Defense of the Modern Synthesis Consensus View. *Philosophy and Theory in Biology* 2. dx.doi.org/10.3998/ptb.6959004.0002.003

Mesoudi, A., S. Blanchet, A. Charmentier, et al. (2013). Is Non-Genetic Inheritance Just a Proximate Mechanism? A Corroboration of the Extended Evolutionary Synthesis. *Biological Theory* 7:189–195.

Mill, J.S. (1843). *A System of Logic Ratiocinative and Inductive.* London: Harper and Brothers. www.Gutenberg.Org/Files/27942/27942-Pdf.Pdf

Millikan, R.G. (1984). *Language, Thought and Other Biological Processes.* Cambridge, MA: MIT Press.

Millikan, R.G. (1989a). Biosemantics. In *White Queen Psychology and Other Essays for Alice.* Cambridge, MA: MIT Press.

Millikan, R.G. (1989b). In Defence of Proper Functions. *Philosophy of Science* 56: 288–302.

Millstein, R. (2006). Natural Selection as a Population-Level Causal Process. *British Journal for the Philosophy of Science* 57: 627–653.

Mitchell, S. (2008). Exporting Causal Knowledge in Evolutionary and Developmental Biology. *Philosophy of Science* **75**: 697–706.

Mitchell, S. (2012). *Emergence: Logical, Functional and Dynamical.* Synthese 185: 171–186.

Moczek, A.P. (2012). The Nature of Nurture and the Future of Evo-Devo: Toward a Theory of Developmental Evolution. *Integrative and Comparative Biology* 52: 108–119.

Moczek, A.P., S. Sultan, S. Foster, C. Ledón-Rettig, I. Dworkin, H.F. Nijhout, E. Abouheif and D.W. Pfennig (2011). The Role of Developmental Plasticity in Evolutionary Innovation. *Proceedings of the Royal Society B.* DOI:10.1098/Rspb.2011.0971

Monod, J. (1971). *Chance and Necessity: An Essay on the Metaphysics of Life* (Trans. A. Wainhouse). New York: Schopf and Sons.

Morange, M. (1998). *A History of Molecular Biology.* Cambridge, MA: Harvard University Press.

Morange, M. (2009). *Life Explained.* New Haven: Yale University Press.

Morange, M. (2011). Evolutionary Developmental Biology: Its Roots and Characteristics. *Developmental Biology* 357: 13–16.

Morgan, T.H. (1925). *Evolution and Genetics.* New York: Columbia University Press.

Morgan, T.H. (1910). Chromosomes and Heredity. *The American Naturalist* 44: 449–496.

Morgan, T.H. (1926). *The Theory of the Gene.* New Haven: Yale University Press.

Morgan, T.H. (1934). The Relation of Genetics to Physiology and Medicine. Nobel Lecture 4 June 1934, pp. 313–328. www.Nobelprize.Org/Nobel_Prizes/Medicine/Laureates/1933/Morgan-Lecture.Pdf

Morrison, M. (2002). Modeling Populations: Pearson and Fisher on Mendelism and Biometry. *British Journal For the Philosophy of Science* 53: 339–368.

Moss, L. (2003). *What Genes Can't Do.* Cambridge, MA: MIT Press.

Müller, G.B. (2007). Evo-Devo: Extending the Evolutionary Synthesis. *Nature Reviews Genetics* 8: 943–949. DOI: 10.1038/Nrg2219

Müller G.B. (2008). EvoDevo as a Discipline. In A. Minelli and G. Fusco (Eds.), *Evolving Pathways: Key Themes in Evolutionary Developmental Biology* (pp. 3–29). Cambridge: Cambridge University Press.

Müller, G.B. (2010). Epigenetic Innovation. In M. Pigliucci and G.B. Muller (Eds.), *The Extended Synthesis* (pp. 307–331). Cambridge, MA: MIT Press.

Müller, G.B., and S. Newman (2005). The Innovation Triad: An Evodevo Agenda. *J. Exp. Zool. (Mol. Dev. Evol.)* 304B: 487–503.

Müller-Wille, S., and H.-J. Rheinberger (2012). *A Cultural History of Heredity.* Chicago: Chicago University Press.

Natural History Museum, London (2014). *Welwitschia Mirabilis.* Eol.Org/Pages/1156 352/Details

Neander, K. (1991). The Teleological Notion of 'Function'. *Australasian Journal of Philosophy* 69: 454–468.

Needham, J. (1934) *A History of Embryology.* Cambridge: Cambridge University Press.

Newman, S. (2003). From Physics to Development: The Evolution of Morphogenetic Mechanisms. In G. Muller and S. Newman (Eds.), *Origination of Organismal Form* (pp. 221–390). Cambridge, MA: MIT Press.

Newman, S. (2011). Complexity in Organismal Evolution. In C. Hooker (Ed.), *Philosophy of Complex Systems* (pp. 335–354). London: Elsevier.

Newman, S. (2014). Form and Function Remixed: Developmental Physiology and the Evolution of Vertebrate Body Plans. *Journal of Physiology* 592.11: 2403–2412.

Newman, S. (2015). Development and Evolution: The Physics Connection. In A. Love (Ed.), *Conceptual Change in Biology: Scientific and Philosophical Perspectives on Evolution and Development* (pp. 421–440). Dordrecht: Springer.

Newman, S., and R. Bhat (2008). Dynamical Patterning Modules: Physico-Genetic Determinants of Morphological Development and Evolution. *Physical Biology* 5 015008 (14pp). DOI: 10.1088/1478–3975/5/1/015008

Newman, S., and R. Bhat (2009). Dynamical Patterning Modules: A 'Pattern Language' for Development and Evolution of Multicellular Form. *International Journal For Developmental Biology* 53: 693–705.

Newman, S., and G.B. Muller (2000). Epigenetic Mechanisms of Character Origination. *Journal of Experimental Zoology* B288: 304–317.

Newman, S., and G.B. Muller (2007). 'Genes and Form'. In E. Neuman-Held and C. Rehman-Suter (Eds.), *Genes in Development: Re-Reading the Molecular Paradigm.* Durham, NC: Duke University Press.

Newman, S.A., G. Forgacs and G.B. Muller (2006). Before Programs: The Physical Origination of Multi-Cellular Forms. *International Journal of Developmental Biology* 50: 289–299.

Newton, I. (1686). *The Mathematical Principles of Natural Philosophy.* New York: Daniel Adler. http://archive.org/stream/newtonspmathema00newtrich#page/n7/mode/2up

Nicholson, D. (2013). Organisms ≠ Machines. *Studies in the History and Philosophy of Biological and Biomedical Sciences* 44: 669–678.

Nicholson. D. (2014). The Machine Conception of the Organism in Development and Evolution: A Critical Analysis. *Studies in History and Philosophy of Biological and Biomedical Sciences* 48: 162–74.

NOAA Fisheries (2013). Marine Mammal Education Web. www.Afsc.Noaa.Gov/N mml/Education/Cetaceans/Blue.Php (accessed 3 September 2015).

Noble, D. (2006). *The Music of Life*. Oxford: Oxford University Press.

Noble, D. (2008). Genes and Causation. *Philosophical Transactions of the Royal Society A* 366: 3001–3015.

Noble, D. (2012). A Theory of Biological Relativity: No Privileged Level of Causation. *interface Focus* 2: 55–64. DOI: 10.1098/Rsfs.2011.0067.

Noble, D. (2013). Evolution beyond Neo-Darwinism. *Journal of Experiment Biology* (Draft Feb 27).

Noble, D., E. Jablonka, M. Joyner, G. Müller and S. Ohmolt (2014). Evolution Evolves: Physiology Returns to Centre Stage. *Journal of Physiology* 592: 2237–2244.

Northcott, R. (2009). Is Actual Difference Making Actual? *Journal of Philosophy* 106: 629–633.

O'Connor, T. (1994). Emergent Properties. *American Philosophical Quarterly* 31: 91–104.

Odling-Smee, F.J. (2010). Niche Inheritance. In M. Pigliucci and G.B. Muller (Eds.) *The Extended Synthesis* (pp. 175–207). Cambridge, MA: MIT Press.

Odling-Smee, F.J., K. Laland and M. Feldman (2003). *Niche Construction: The Neglected Process in Evolution*. Princeton: Princeton University Press.

Okasha, S. (2006). *Evolution and the Levels of Selection*. Oxford: Oxford University Press.

Okasha, S. (2008). Fisher's Fundamental Theorem of Natural Selection – A Philosophical Analysis. *British Journal For the Philosophy of Science* 59: 319–351.

Oppenheim, P., and H. Putnam (1958). The Unity of Science as a Working Hypothesis. In H. Feigl et al. (Eds.), *Minnesota Studies in the Philosophy of Science*, Vol. 2. Minneapolis: Minnesota University Press.

Oppenheimer, M.J. (1967). Two Puzzles for the Origin of Species. In M.J. Oppenheimer, *Essays in the History of Embryology and Biology*. Cambridge, MA: MIT Press.

Orr, H.A. (2005). The Genetic Theory of Adaptation: A Brief History. *Nature Reviews Genetics* 6: 119–127.

Orr, H.A. (2007a). Theories of Adaptation: What They Do and Don't Say. *Genetica* 123: 3–13.

Orr, H.A. (2007b). Absolute Fitness, Relative Fitness, and Utility. *Evolution* 61: 2997–3000.

Orr, H.A., and J.A. Coyne (1992). The Genetics of Adaptation Revisited. *American Naturalist*. 140: 725–742.

Orzack S., and E. Sober (1994). How to Formulate and Test Adaptationism. *American Naturalist* 148: 202–210.

Otsuka, J., T.Y. Turner, C. Allen and E. Lloyd (2011). Why the Causal View of Fitness Survives. *Philosophy of Science*. 78: 209–224.

Ou, X., Y Zhang, C. Xu, X. Lin, Q. Zang, et al. (2012). Transgenerational Inheritance of Modified DNA Methylation Patterns and Enhanced Tolerance Induced by Heavy Metal Stress in Rice (*Oryza Sativa* L.). *Plos ONE* 7 (9): E41143. DOI: 10.1371/Journal.Pone.0041143.

Oyama, S. (1985). *The Ontogeny of Information*. Durham, NC: Duke University Press.

Oyama, S. (2000). *Evolution's Eye*. Durham, NC: Duke University Press.

Oyama, S., P.E. Griffiths and R. Gray (Eds.) (2001). *Cycles of Contingency*. Cambridge, MA: MIT Press.

Paley, W. (1809 [2006]). *Natural Theology*. Oxford: Oxford University Press.

Pearce, T. (2009). 'A Great Complication of Circumstances' – Darwin and the Economy of Nature. *Journal of the History of Biology* 43: 493–528. DOI: 10.1007/S10739 -009-9205-0

Pearce, T. (2010). From 'Circumstances' to 'Environment': Herbert Spencer and the Origins of the Idea of Organism-Environment Interaction. *Studies in History and Philosophy of Biological and Biomedical Sciences* 41: 241–252.

Pearce, T. (2011). Evolution and Constraints on Variation: Variant Specification and Range of Assessment. *Philosophy of Science* 78: 739–751.

Pfennig, D.W., M.A. Wund, E.C. Snell-Rood, T. Cruickshank, S. Ciliberti, O.C. Martin and A. Wagner (2007). Innovation and Robustness in Complex Regulatory Gene Networks. *PNAS* 104(34): 13591–13596.

Pfennig, D.W., M.A. Wund, C. Schlichting, E.C. Snell-Rood, T. Cruikshank, C. Schichting and A. Moczek (2010). Phenotypic Plasticity's Impacts on Diversification and Speciation. *Trends in Ecology and Evolution* 25: 459–467.

Pigliucci, M. (2009a). An Extended Synthesis for Evolutionary Biology. The Year in Evolutionary Biology 2009: *Ann. N.Y. Acad. Sci.* 1168: 218–228. DOI: 10.1111/ J.1749–6632.2009.04578.X.

Pigliucci, M. (2009b). An Extended Synthesis for Evolutionary Biology. *Annals of the New York Academy of Sciences* 1168 (pp. 218–228). The Year in Evolutionary Biology.

Pigliucci, M. (2010). Genotype-Phenotype Mapping and the End of the 'Genes As Blueprint' Metaphor. *Philosophical Transactions Royal Society B* 365: 557–566.

Pigliucci, M., and G. Muller (Eds.) (2010). *Evolution: The Extended Synthesis*. Cambridge, MA: MIT Press.

Prigogine, I., and I. Stengers (1984). *Order Out of Chaos*. London: Bantam.

Provine, W.B. (1971). *The Origins of Theoretical Population Genetics*. Chicago: University of Chicago Press.

Purcell, E.M. (1977). Life at Low Reynolds Number. *American Journal of Physics* 45: 101–111.

Putnam, H. (1970). Is Semantics Possible? *Metaphilosophy* 1: 187–201.

Quarfoord, M. (2006). Kant on Biological Teleology: Towards a Two-Level Interpretation. *Studies in History and Philosophy of Biological and Biomedical Sciences* 37: 735–747.

Radick, G. (2012). Should 'Heredity' and 'Inheritance' Be Scientific Terms? William Bateson's Change of Mind as a Historical and Scientific Problem. *Philosophy of Science* 79: 714–724.

Raff, R. (1996). *The Shape of Life: Genes, Development and the Evolution of Animal Form*. Chicago: Chicago University Press.

Richards, E.J. (2006). Inherited Epigenetic Variation – Revisiting Soft Inheritance. *Nature Reviews Genetics*, pp. 395–400.

Rohner, N., et al. (2013). Cryptic Variation in Morphological Evolution: HSP90 As a Capacitor for Loss of Eyes in Cavefish. *Science* 342: 1372–1375. DOI:10.1126/ Science.1240276

Rolian, C., and B. Hallgrìmsson (2009). Integration and Evolvability in Primate Hands and Feet. *Evolutionary Biology*. 36: 100–117.

Rolian, C., D.E. Lieberman and B. Hallgrìmsson (2010). The Co-Evolution of Hands and Feet. *Evolution* 64: 1558–1568.

Roll-Hansen, N. (2009). Sources of Wilhelm Johannsen's Genotype Theory. *Journal of the History of Biology* 42: 457–493.

Rosenberg, A. (2006). *Darwinian Reductionism: Or, How to Stop Worrying and Love Molecular Biology*. Chicago: University of Chicago Press.

Rosenbleuth, A., N. Wiener and J. Bigelow (1943). Behavior, Purpose and Teleology. *Philosophy of Science* 10: 18–24.

Roux. S. (2005). Empedocles to Darwin. In E. Close, M. Tsianikas and G. Frazis (Eds.). *Greek Research in Australia: Proceedings of the Biennial international Conference of Greek Studies, Flinders University April 2003* (pp. 1–16). Flinders University Department of Languages – Modern Greek: Adelaide.

Ruiz-Mirazoa, K.J., and A. Moreno (2004). A Universal Definition of Life. *Origins of Life and Evolution of the Biosphere* 34: 323–346.

Ruiz-Mirazo, K.J., and A. Moreno (2012). Autonomy in Evolution: From Minimal to Complex Life. *Synthese* 185: 21–52.

Ruiz-Mirazo, K., et al. (2000). Organisms and Their Place in Biology. *Theory in Biosciences* 119: 209–233.

Ruse, M. (1971). Functional Statements in Biology. *Philosophy of Science* 38: 87–95.

Ruse, M. (1989). Do Organisms Exist? *American Zoologist* 29: 1061–1066.

Ruse, M. (2003). *Darwin and Design: Does Evolution Have a Purpose?* Cambridge, MA: Harvard University Press.

Russell E.S. (1945). *The Directiveness of Organic Activities*. Cambridge: Cambridge University Press.

Russert-Kraemer, L., and W.J. Bock (1989). Prologue: The Necessity of Organisms. *American Zoologist* 29: 1056–1060.

Rutherford, S.L., and S. Lindquist (1998). Hsp90 as a Capacitor for Morphological Evolution. *Nature* 396: 336–342. DOI:10.1038/24550

Salmon, W. (1984). *Scientific Explanation and the Causal Structure of the World*. Princeton: Princeton University Press.

Sanders, J.T. (1993). Merleau-Ponty, Gibson, and the Materiality of Meaning. *Man and World* 26: 287–302.

Sansom, R. (2009). Evolvability. In M. Ruse (Ed.), *The Oxford Handbook of Philosophy of Biology* (pp. 138–160). Oxford: Oxford University Press.

Sansom, R. (2011). *Ingenious Genes: How Gene Regulation Networks Evolve to Control Development*. Cambridge, MA: MIT Press.

Sarkar, S. (1996). Biological Information: A Sceptical Look at Some Central Dogmas of Molecular Biology. In S. Sarkar (Ed.), *The Philosophy and History of Molecular Biology: New Perspectives* (pp. 187–232). Dordrecht: Kluwer Academic Publishers.

Schaffner, K. (1969). The Watson-Crick Model and Reductionism. *British Journal for the Philosophy of Science* 20: 325–348.

Schlichting, C.D. (2003). The Origins of Differentiation via Phenotypic Plasticity. *Evolution and Development* 5: 98–105.

Schlichting, C.D., and A.P. Moczek (2010). Phenotypic Plasticity's Impact on Diversification and Speciation. *Trends in Ecology and Evolution* 25: 459–467.

Schlick, M. (1953). Philosophy of Organic Life. In H. Feigl and M. Brodbeck (Eds.), *Readings in the Philosophy of Science* (pp. 523–536). New York: Appleton-Century Crofts.

Schlosser, G. (2002). Modularity and the Units of Evolution. *Theory Biosciences* 121: 1–80.

Schmalhausen, I.I. (1948 [1986]). *Factors of Evolution: The Theory of Stabilizing Selection*. Chicago: Chicago University Press.

Schmitz, R.J., et al. (2013). Patterns of Population Epigenomic Diversity. *Nature* 495: 193–198. DOI: 10.1038/Nature11968

Schneider E.D., and D. Sagan (2007). *Into the Cool*. Chicago: Chicago University Press.

Schneider, E.D., and J.J. Kay (1995). Order from Disorder: The Thermodynamics of Complexity in Biology. In M.P. Murphy and L.A.J. O'Neill (Eds.), *What Is Life? The Next Fifty Years* (pp. 161–174).Cambridge: Cambridge University Press.

Schrödinger, E. (1944). *What Is Life?* New York: Dover.

Schwenk, K., and G. Wagner (2001). Function and the Evolution of Phenotypic Stability: Connecting Pattern to Process. *American Zoologist* 41: 522–563.

Schwenk, K., and G. Wagner (2004), The Relativism of Constraints on Phenotypic Evolution. In M. Pigliucci and K. Preston (Eds.), *The Evolution of Complex Phenotypes* (pp. 390–408). Oxford: Oxford University Press.

Shapin, S. (1996). *The Scientific Revolution*. Chicago: University of Chicago Press.

Shapiro, J.A. (2007). Bacteria Are Small But Not Stupid. Cognition, Natural Genetic Engineering and Socio-Bacteriology. *Studies in History and Philosophy of Biological and Biomedical Sciences* 38: 807–819. DOI: 10.1016/J.Shpsc.2007.09.010.

Shapiro, J.A. (2011). *Evolution: A View from the 21st Century Perspective*. Upper Saddle River, NJ: FT Press Science.

Shapiro, J.A. (2013). How Life Changes Itself: The Read-Write (RW) Genome. *Physics of Life Reviews* 10: 287–323.

Shapiro, J.A. (2014). Physiology of the Read-Write Genome. *Journal of Physiology* 592.11: 2319–2341.

Shapiro, L. (2014). *Mastophora* Bolas Spiders. *Encyclopedia of Life*. Eol.Org/Pages/112760/Details

Shea, N. (2007). Representation in the Genome and in Other Inheritance Systems. *Biology and Philosophy* 22: 313–331.

Shea, N. (2011). Developmental Systems Theory Formulated as a Claim about Inherited Information. *Philosophy of Science* 78: 60–82.

Shields, C. (2014). *Aristotle* (2nd ed.). London: Routledge.

Shubin, M., C. Tabin and S. Carroll (2009). Deep Homology and the Origins of Evolutionary Novelty. *Nature* 457:818–823. DOI: 10.1038/Nature07891

Simpson, G.G. (1944). *Tempo and Mode in Evolution*. New York: Columbia University Press.

Simpson, G.G. (1953). The Baldwin Effect. *Evolution* 7: 110–117.

Singer M.C., and C.D. Thomas (1996). Evolutionary Responses of a Butterfly Metapopulation to Human and Climate-Caused Environmental Variation. *American Naturalist* 148: S9–S39.

Sloan, P. (1977). Descartes, the Sceptics, and the Rejection of Vitalism in Seventeenth-Century. *Physiology Studies in the History and Philosophy of Science* 8: 2–27.

Smith, M.L., J.N. Bruhn and J.B. Anderson. (1992). The Fungus *Armillaria Bulbosa* Is Among the Largest and Oldest Living Organisms. *Nature* 356: 428–431. DOI: 10.1038/356428a0

Smolin, L. (2013). *Time Reborn: From the Crisis in Physics to the Future of the Universe*. New York: Houghton Mifflin Harcourt.

Sober, E. (1980). Evolution, Population Thinking and Essentialism. *Philosophy of Science* 47: 350–383.

Sober, E. (1984). *The Nature of Selection*. Cambridge, MA: MIT Press.

Sober, E. (1987). A Plea for Pseudoprocesses. *Pacific Philosophical Quarterly* 66: 303–309.

Sober, E. (2006). Evolution, Population Thinking and Essentialism. In E. Sober (Ed.), *Conceptual Issues in Evolutionary Biology* (pp. 329–359). Cambridge, MA: MIT Press (Originally Published in 1980).

Sober, E. (2013). Trait Fitness Is Not a Propensity, But Fitness Variation Is. *Studies in the History and Philosophy of Biological and the Biomedical Sciences* 44: 336–341.

Sommerhoff, G. (1950). *Systems Biology*. Cambridge: Cambridge University Press.

Sorabji, R. (1990). *Necessity, Cause, and Blame*. London: Duckworth.

Stanley, S. (1998). *Macroevolution: Pattern and Process*. Baltimore: Johns Hopkins University Press.

Stegenga, J. (2014). Population Pluralism and Natural Selection. *British Journal for the Philosophy of Science*. DOI: 10.1093/Bjps/Axu003

Stegmann, U. (2005). Genetic information as Instructional Content. *Philosophy of Science* 72: 425–443.

Stegmann, U. (2012). Varieties of Parity. *Biology and Philosophy* 27: 903–918.

Stephens, C. (2004). Selection, Drift, and the 'Forces' of Evolution. *Philosophy of Science* 71: 550–570.

Sterelny, K. (2000a). The 'Genetic Program' Program: A Commentary on Maynard Smith on Information in Biology. *Philosophy of Science* 67 (2): 195–201.

Sterelny, K. (2000b). Development, Evolution, and Adaptation. *Philosophy of Science* 67(Supplement): S369–S387.

Sterelny, K. (2001). Niche Construction, Developmental Systems, and the extended Replicator. In S. Oyama, P.E. Griffiths and R.D. Gray (Eds.), *Cycles of Contingency: Developmental Systems and Evolution* (pp. 331–349). Cambridge, MA: MIT Press.

Sterelny, K. (2005). Made by Each Other: Organisms and Their Environment. *Biology and Philosophy* 20: 21–36.

Sterelny, K. (2007). What Is Evolvability? In M. Matthen and C. Stephens (Eds.), *Handbook of the Philosophy of Science Philosophy of Biology* (pp. 163–178). Amsterdam: Elsevier.

Sterelny, K. (2009). Novelty, Plasticity and Niche Construction: The Influence of Phenotypic Variation on Evolution. In A. Barberousse, M. Morange and T. Pradeu (Eds.), *Mapping the Future of Biology: Evolving Concepts and Theories* (pp. 93–110). Dordrecht: Springer.

Sterelny, K., and P.E. Griffiths (1999). *Sex and Death: An Introduction to Philosophy of Biology*. Chicago: Chicago University Press.

Stern, D.L. (2000). Evolutionary Developmental Biology and the Problem of Variation. *Evolution* 54: 1079–1091.

Stoffregen, T. (2003). Affordances as Properties of the Animal-Environment System. *Ecological Psychology*, 15: 115–134.

Stossel, T. (1999). Crawling Neutrophil Chasing a Bacterium. https://embryology.m ed.unsw.edu.au/embryology/index.php/Movie_-_Neutrophil_ chasing_bacteria (accessed 3 September 2015).

Stotz, K. (2006a). 'With Genes Like That, Who Needs an Environment? Postgenomics's Argument for the 'Ontogeny of information'. *Philosophy of Science* 73: 905–917.

Stotz, K. (2006b). Molecular Epigenesis: Distributed Specificity as a Break in the Central Dogma. *History and Philosophy of the Life Sciences* 26: 527–544.

Stout, R. (1996). *Things That Happen Because They Should*. Oxford: Oxford University Press.

Strevens, M. (2005). How Are the Sciences of Complex Systems Possible? *Philosophy of Science* 72: 531–556.

Szathmáry, E. (2000). The Evolution of Replicators. *Philosophical Transactions of the Royal Society of London. Series B: Biological Sciences*, 355: 1669–1676.

Talbot, S.D. (2013). The Myth of the Machine-Organism. In S. Krimsky and J. Gruber (Eds.), *Genetic Explanations: Sense and Nonsense* (pp. 52–68). Cambridge, MA: Harvard University Press.

Taylor, C. (1963). *The Explanation of Behaviour*. London: Routledge.

Taylor, C. (2005). Merleau-Ponty and the Epistemological Picture. In T. Carman and M.B. Hansen (Eds.), *The Cambridge Companion to Merleau-Ponty* (pp. 26–49). Cambridge: Cambridge University Press.

Tennant, N. (2014). The Logical Structure of Evolutionary Explanation and Prediction: Darwinism's Fundamental Schema. *Biology and Philosophy*. April 2014. DOI: 10.1007/S10539-014-9444-0

Tero, A., T. Nakagaki and Y. Kuramoto (2008). Amoebae Anticipate Periodic Events. *Physical Review Letters* 100: 018101-1–018101-4.

Teru, A., S. Takagi, T. Saigusa, et al. (2010). Rules for Biologically Inspired Adaptive Network Design. *Science* 327: 439–442.

Thom, R. (1972a). Structuralism and Biology. In C.H. Waddington, *Towards A theoretical Biology*. Vol. 4. Chicago: Aldine Publishing, pp. 82–89.

Thom, R. (1972b). *Structural Stability and Morphogenesis: An Outline of a General Theory of Models* (Trans. D.H. Fowler). Don Mills, Ontario: Addison-Wesley.

Thompson, D.W. (1961). *On Growth and Form* (abridged edition). J.T. Bonner (Ed.). Cambridge: Cambridge University Press.

Thompson, E. (2007). *Mind in Life: Biology, Phenomenology and the Sciences of Mind*. Cambridge, MA: Harvard University Press.

Trinkaus, J.P. (1984). *Cells into Organs* (2nd ed.). Edgewood Cliffs, NJ: Prentice Hall.

True, R., and S. Carroll (2002). Gene Co-Option in Physiological and Morphological Evolution. *Annual Reviews of Cell and Developmental Biology* 18: 53–80.

Turner, J.S. (2000). *The Extended Organism: The Physiology of Animal-Built Structures* Cambridge, MA: Harvard University Press.

Turvey. M.T. (1992). Affordances and Prospective Control: An Outline of the Ontology. *Ecological Psychology* 4: 173–187.

Van Duin, M., F. Keijzer and D. Franken (2006). Principles of Minima Cognition: Casting Cognition as Sensorimotor Co-ordination. *Animats, Software Agents, Robots, Adaptive Systems* 14: 157–170.

Vane-Wright, D. (2011). Whatever Happened to the Organic Selectionists? *Antenna*, Chiswell Green 35(2): 57–60.

Vane-Wright, D. (2014). What Is Life? and What Might Be Said of the Role of Behaviour in Its Evolution? *Biological Journal of the Linnean Society* 112: 219–241.

Varela, F.H., H. Maturana and R. Uribe (1974). Autopoiesis: The Organization of Living Systems, Its Characterization and a Model. *Biosystems* 5: 187–196.

Varela, F., E. Thompson and E. Rosch (1991). *The Embodied Mind: Cognitive Science and Human Experience.* Cambridge, MA: MIT Press.

Velleman, D.L. (1992). What Happens When Someone Acts? *Mind* 101 (403): 461–481.

von Bertalanffy, L. (1950). The Theory of Open Systems in Physics and Biology. *Science* 111: 23–9.

von Bertalanffy, L. (1969). *General Systems Theory.* New York: George Braziller.

Von Dassow, G., and E.M. Munro (1999). Modularity in Animal Development and Evolution: Elements of a Framework for Evo-Devo. *Journal of Experimental Zoology (Mol. Dev. Evol.)* 285: 307–325.

Von Dassow, G., E. Meir, E.M. Munro and G.M. Odell (2000). The Segment Polarity Network Is a Robust Developmental Module. *Nature* 406: 188–192.

von Helmholtz, H. (1971). The Aim and Progress of Physical Science. Reprinted in R. Kahl (Ed.). *Selected Writings of Hermann Von Helmholtz* (pp. 223–245). Middletown, CT: Wesleyan University Press.

Von Uexküll, J. (1938). Die Neue Umweltlehre: Ein Bindeglied Zwischen Natur-Und Kulturwissenschaften. *Die Ehrziehung* 13: 185–199.

Waddington, C.H. (1956). Genetic Assimilation of the Bithorax Phenotype. *Evolution* 10: 1–13.

Waddington, C.H. (1957). *The Strategy of the Genes.* London: Allen and Unwin.

Waddington, C.H. (1960). Evolutionary Adaptation. In S. Tax (Ed.), *Evolution after Darwin.* Chicago: University of Chicago Press, 381–402.

Waddington, C.H. (1966). *Principles of Development and Differentiation.* London: Macmillan.

Wagner, A. (1996). Can Nonlinear Epigenetic Interactions Obscure Causal Relations between Genotype and Phenotype? *Nonlinearity* 9: 607–629.

Wagner, A. (1999). Causality in Complex Systems. *Biology and Philosophy* 14: 83–101.

Wagner, A. (2005). *Robustness and Evolvability in Living Systems.* Princeton, NJ: Princeton University Press.

Wagner, A. (2007). Distributional Robustness versus Redundancy as Causes of Mutational Robustness. *Bioessays* 27: 176–188.

Wagner, A. (2011). *The Origin of Evolutionary Innovations: A Theory of Transformative Change in Living Systems.* Oxford: Oxford University Press.

Wagner, A. (2012). The Role of Robustness in Phenotypic Adaptation and Innovation *Proceedings of the Royal Society B* 279: 1249–1258.

Wagner, A. (2014). *The Arrival of the Fittest: Solving Evolution's Greatest Puzzle.* New York: Current Books.

Wagner, G. (2000). What Is the Promise of Developmental Evolution? Part I: Why Is Developmental Biology Necessary to Explain Evolutionary Innovations. *Journal of Experimental Zoology (Mol Dev Evol)* 288: 95–98.

Wagner, G., and L. Altenberg (1996). Complex Adaptations and the Evolution of Evolvability. *Evolution* 50: 967–976.

Wallace, B. (1986). Can Embryologists Contribute to an Understanding of Evolutionary Mechanisms?. In W. Bechtel (Ed.), *Integrating Scientific Disciplines* (pp. 149–163). Dordrecht: M. Nijhoff.

Walsh, D.M. (1999). Alternative Individualism. *Philosophy of Science*. 66: 628–48.

Walsh, D.M. (2000). Chasing Shadows. *Studies in History and Philosophy of Biological and Biomedical Sciences* 31: 135–153.

Walsh, D.M. (2003). Fit and Diversity: Explaining Adaptive Evolution. *Philosophy of Science* 70: 280–301.

Walsh, D.M. (2004). Bookkeeping or Metaphysics? The Units of Selection Debate. *Synthese* 138: 337–361.

Walsh, D.M. (2006). Evolutionary Essentialism. *British Journal for the Philosophy of Science* 57: 425–448.

Walsh, D.M. (2007a). The Pomp of Superfluous Causes. *British Journal for the Philosophy of Science* 74: 281–303.

Walsh, D.M. (2007b). Development: Three Grades of Ontogenetic Involvement. In M. Matthen and C. Stephens (Eds.), *Handbook of the Philosophy of Science* (pp. 179–199). Amsterdam: North-Holland.

Walsh, D.M. (2007c). Organisms As Natural Purpose: The Contemporary Evolutionary Perspective. *Studies in the History and Philosophy of Biological and Biomedical Sciences* 37: 771–791.

Walsh, D.M. (2008). Teleology. In M. Ruse (Ed.), *Oxford Handbook of the Philosophy of Biology* (pp. 113–137). Oxford: Oxford University Press.

Walsh, D.M. (2010a). Not A Sure Thing. *Philosophy of Science* 77: 147–71.

Walsh, D.M. (2010b). Two Neo-Darwinisms. *History and Philosophy of the Llife Sciences* 32: 317–339.

Walsh, D.M. (2012a). Situated Adaptationism. In W. Kabesenche, M. O'Rourke and M. Slater (Eds.), *The Environment: Philosophy, Science, Ethics* (pp. 89–116). Cambridge, MA: MIT Press.

Walsh, D.M. (2012b). Mechanism and Purpose: A Case for Natural Teleology. *Studies in the History and Philosophy of Biology and the Biomedical Sciences* 43:173–181.

Walsh, D.M. (2013a). Adaptation and the Affordance Landscape: The Spatial Metaphors of Evolution. In G. Barker, E. Desjardins and T. Pearce (Eds.), *Entangled Life* (pp. 213–36). Dordrecht: Springer.

Walsh, D.M. (2013b). Mechanism, Emergence, and Miscibility: The Autonomy of Evo-Devo. In P. Huneman (Ed.), Functions: Selection and Mechanisms. *Synthese Library* 363: 43–65.

Walsh, D.M. (2014a). Variance, Invariance, and Statistical Explanation. *Erkkenntnis*. DOI: 10.1007/S10670-014–9680-3

Walsh, D.M. (2014b). The Negotiated Organism: Inheritance, Development and the Method of Difference. *Biological Journal of the Linnean Society* 112: 295–30.

Walsh, D.M. (2014c). Function and Teleology. In R.P. Thompson and D.M. Walsh (Eds.), *Evolutionary Biology: Conceptual, Ethical, and Religious Issues*. Cambridge: Cambridge University Press.

Walsh, D.M. (2014d). Descriptions and Models: Some Replies to Abrams. *Studies in the History and Philosophy of Biological and the Biomedical Sciences* 44: 302–308.

Walsh, D.M. (Forthcoming). Chance Caught on the Wing: Metaphysical Commitment or Methodological Artifact?. In P. Huneman and D.M. Walsh (Eds.), *Challenges to Evolutionary Theory*. Oxford: Oxford University Press.

Walsh, D.M., T. Lewens, and A. Ariew (2002). The Trials of Life. *Philosophy of Science*. 69: 452–473.

Waters, C.K. (1994). Genes Made Molecular. *Philosophy of Science*. 61: 163–185.

Waters, K. (2007). Causes That Make a Difference. *Journal of Philosophy* 104: 551–579.

Watson, J. (1965). *The Molecular Biology of the Gene*. New York: W.A. Benjamin.

Watson, J., and F. Crick (1953). Molecular Structure of Nucleic Acids. *Nature* (25April): 737.

Weaver, I.C.T., N. Cervoni, F.F. Champagne, A.C.V. D'Alesssio, S. Sharma, J.R. Seckl, S. Dymov, M. Szyf and M.J. Meamey (2004). Epigenetic Programming by Maternal Behavior. *Nature Neuroscience* 7:847–854.

Weaver, W. (1948). Evolution and Complexity. *Scientific American*. 36: 536–544.

Weber, B., and D. Depew (2003). *Evolution and Learning: The Baldwin Effect Reconsidered*. Cambridge: Cambridge University Press.

Weber, M. (Forthcoming). Causal Selection vs. Causal Parity in Biology: Relevant Counterfactuals and Biologically Normal Interventions. In C.K. Waters, M. Travisano, and J. Woodward (Eds.), *Philosophical Perspectives on Causal Reasoning in Biology*. Minnesota Studies in the Philosophy of Science.

Webster, G., and B. Goodwin (1986). *Form Ant Transformation: Generative and Relational Principles in Biology*. Cambridge: Cambridge University Press.

Weisberg, M. (2006). Forty Years of 'The Strategy': Levins on Model Building and Idealization. *Philosophy and Biology* 21:526–645.

West-Eberhard, M.J. (2003). *Developmental Plasticity and Evolution*. Oxford: Oxford University Press.

West-Eberhard, M.J. (2005a). Developmental Plasticity and the Origin of Species Differences. *Proceedings of the National Academy of Sciences* 102: 6543–6349.

West-Eberhard, M.J. (2005b). Phenotypic Accommodation: Adaptive Innovation Due to Developmental Plasticity. *Journal Experimental Zoology (Mole Dev Evo)* 304B: 610–618.

Wilkins, A. (2011). Why Did the Modern Synthesis Give Short Shrift to 'Soft inheritance'. In B. Snait and E. Jablonka (Eds.), *Transformations of Lamarckism: From Subtle Fluids to Molecular Biology* (pp. 127–132). Cambridge, MA: MIT Press.

Williams, G.C. (1966). *Adaptation and Natural Selection: A Critique of Some Current Evolutionary Thought*. Princeton: Princeton University Press.

Willmer, P., G. Stone and I. Johnstone (2005). *Environmental Physiology of Animals*, 2nd ed. Oxford: Blackwell Publishing.

Wilson, E.O. (1980). *Sociobiology: The New Synthesis* (abridged edition). Cambridge, MA: Harvard, Belknap Press.

Wimsatt, W. (1974). Reductive Explanation: A Functional Account. *PSA: Proceedings of the Biennial Meeting of the Philosophy of Science Association* 1974: 671–710.

Wimsatt, W. (2000). Emergence as Non-Aggregativity and the Biases of Reductionism. *Foundations of Science* 5: 269–297. Reprinted in W. Wimsatt

(2007), *Re-Engineering Philosophy of Limited Beings* (pp. 274–312). Cambridge, MA: Harvard University Press.

Winther, G.G. (2001). August Weismann on Germ-Plasm Variation. *Journal of the History of Biology* 34: 517–555.

Winther, R. (2011). Part-Whole Science. *Synthese* 178: 397–427.

Winther, R., R. Giordano, M. Edge and R. Nielsen (2015). The Mind, the Lab, and the Field, Three Kinds of Populations in Scientific Practice. *Studies in History and Philosophy of Science Part C: Studies in History and Philosophy of Biological and Biomedical Sciences.*

Wolfe, C. (2010). Do Organisms Have an Ontological Status?' *History and Philosophy of the Life Sciences* 32: 195–232.

Wolfe, C. (2011). From Substantival to Functional Vitalism and Beyond. *Eidos* 14: 212–235.

Woodward, J. (2000). Explanation and Invariance in the Special Sciences. *British Journal for the Philosophy of Science* 51: 107–254.

Woodward, J. (2003). *Making Things Happen.* Oxford: Oxford University Press.

Wray, G.A., H.E. Hoekster, D.J. Futuyma, R.E. Lenski, T.F.C. Mackay, D. Schluter and J.E. Strassman, J.E. (2014). Does Evolutionary Theory Need a Rethink? Yes: Urgently. *Nature* 514: 161–164.

Wright, S. (1932). The Roles of Mutation, Inbreeding, Crossbreeding and Selection in Evolution. *Proceedings of the Sixth International Congress of Genetics*, pp. 356–366.

Zammito, J. (1991). *The Genesis of Kant's 'Critique of Judgment'.* Chicago: University of Chicago Press.

Zinkernagel, R.F. (2001). Maternal Antibodies, Childhood Infections, and Autoimmune Diseases. *New England Journal of Medicine* 345: 1331–1335.

Zollman, K. (2011). Separating Directives and Assertions in Signaling Games. *Journal of Philosophy* 108: 158–169.

Index

Printed in the United States
By Bookmasters